RADIATION INACTIVATION OF BIOTERRORISM AGENTS

NATO Science Series

A series presenting the results of scientific meetings supported under the NATO Science Programme.

The series is published by IOS Press and Springer Science and Business Media in conjunction with the NATO Public Diplomacy Division.

Sub-Series

I.	Life and Behavioural Sciences	IOS Press
II.	Mathematics, Physics and Chemistry	Springer Science and Business Media
III.	Computer and Systems Sciences	IOS Press
IV.	Earth and Environmental Sciences	Springer Science and Business Media
V.	Science and Technology Policy	IOS Press

The NATO Science Series continues the series of books published formerly as the NATO ASI Series.

The NATO Science Programme offers support for collaboration in civil science between scientists of countries of the Euro-Atlantic Partnership Council. The types of scientific meeting generally supported are "Advanced Study Institutes" and "Advanced Research Workshops", although other types of meeting are supported from time to time. The NATO Science Series collects together the results of these meetings. The meetings are co-organized by scientists from NATO countries and scientists from NATO's Partner countries – countries of the CIS and Central and Eastern Europe.

Advanced Study Institutes are high-level tutorial courses offering in-depth study of latest advances in a field.
Advanced Research Workshops are expert meetings aimed at critical assessment of a field, and identification of directions for future action.

As a consequence of the restructuring of the NATO Science Programme in 1999, the NATO Science Series has been re-organized and there are currently five sub-series as noted above. Please consult the following web sites for information on previous volumes published in the series, as well as details of earlier sub-series:

http://www.nato.int/science
http://www.springeronline.nl
http://www.iospress.nl
http://www.wtv-books.de/nato_pco.htm

Series I: Life and Behavioural Sciences - Vol. 365 ISSN: 1566-7693

Radiation Inactivation of Bioterrorism Agents

Edited by

L.G. Gazsó

National Center for Public Health, National Research Institute for Radiobiology and Radiohygiene, Budapest, Hungary

and

C.C. Ponta

"Horia Hulubei" National Institute for Physics and Nuclear Engineering, Bucharest-Magurele, Romania

IOS Press

Amsterdam • Berlin • Oxford • Tokyo • Washington, DC

Published in cooperation with NATO Public Diplomacy Division

Proceedings of the NATO Advanced Research Workshop on
Radiation Inactivation of Bioterrorism Agents
7–9 March 2004
Budapest, Hungary

ISBN 1 58603 488 X
Library of Congress Control Number: 2004116329

Publisher
IOS Press
Nieuwe Hemweg 6B
1013 BG Amsterdam
Netherlands
fax: +31 20 620 3419
e-mail: order@iospress.nl

Distributor in the UK and Ireland
IOS Press/Lavis Marketing
73 Lime Walk
Headington
Oxford OX3 7AD
England
fax: +44 1865 750079

Distributor in the USA and Canada
IOS Press, Inc.
4502 Rachael Manor Drive
Fairfax, VA 22032
USA
fax: +1 703 323 3668
e-mail: iosbooks@iospress.com

PRINTED IN THE NETHERLANDS

Foreword

R. Joel LOWY[a], Thomas B. ELLIOTT[a], Michael O. SHOEMAKER[a], Gregory B. KNUDSON[a] and Marc F. DESROSIERS[b]
[a]*Armed Forces Radiobiology Research Institute, Bethesda, Maryland, USA*
[b]*Ionizing Radiation Division, Physics Laboratory, National Institute of Standards and Technology, Gaithersburg, Maryland, USA*

The use of and problems associated with biological weapons have been of concern to NATO and non-NATO military organizations for many years. Until recently most of the readily available literature addressed the military issues associated with the possible use of biological weapons on the battlefield, the medical effects of the various agents, and what was known about medical prophylaxis and treatments. Information on other needed countermeasures, such as decontamination or public health issues associated with exposure of civilian populations were largely overlooked. This perspective changed dramatically after the events in the United States in the fall of 2001 in response to the contamination of the U.S. Mail system with powdered anthrax spores. Surprisingly, this use of a biological warfare agent, which gained widespread international attention, was not in the context of a military operation, but a terrorist action that involved civilians. The contaminated facilities were both private and government owned, including important mail-sorting and government offices within the Washington DC area. The exposed populations, including those killed by the attack, were exclusively civilians, and not those whose responsibilities would previously have been expected to put them at risk. Most likely, the intensity of the situation, sense of vulnerability of national security, and urgency to provide solutions were magnified by the context of the other unprecedented terrorist attacks that had occurred a few weeks earlier in the United States on 11 September.

Among the significant problems and defensive weaknesses that the anthrax attack revealed was the lack of established industrial-scale decontamination methods for large volumes of heterogeneous objects (e.g., the mail) or for complex physical environments, (e.g., the U.S. Postal Service sorting facilities). Ultimately, these two microbial decontamination problems were solved in very different ways. The contaminated mail was treated with ionizing radiation while the contaminated government buildings were treated with vapor/gas-phase chemicals. The most urgent problems at the time, decontaminating the mail and establishing a process for prophylactic treatment of the mail, were solved relatively quickly. This was largely due to the robust radiation biology and technical base derived from the industrial use of ionizing radiation. For many years this industry has successfully used radiation to sterilize complex objects, and indeed most modern hospitals are wholly dependent on the wide variety of medical devices and supplies sterilized by such methods. Contributing to the speed of response was the fact that the attack occurred within the United States and in the "home town" of many of the technical experts and decision makers, allowing official response to be coordinated rapidly.

This volume presents the papers delivered at the NATO Advanced Research Workshop on Radiation Inactivation of Bioterrorism Agents, 7-9 March 2004 in Budapest, Hungary. The conference was graciously hosted by the Frédéric Joliot-Curie National Research Institute for Radiobiology and Radiohygiene, Budapest, Hungary and organized by Dr. L.G.

Gazsó and Dr. C.C. Ponta. The conference was in part the outcome of the co-organizers' forward thinking in this area and previous recommendations on the use of ionizing radiation for biological weapon agent inactivation (VI. Int. Symposium on Protection Against Chemical and Biological Warfare Agents, Stockholm, 1998 and Symposium on Nuclear, Biological and Chemical Threats in the 21st Century, Helsinki, 2000). Wisely, the conference brought together experts from across a number of professional disciplines and geographic boundaries from the private sector, government, scientific research, and international regulatory agencies.

The conference papers within this volume cover many of the factors essential to the successful application of ionizing radiation to biological agent inactivation. Consideration of international law and treaty issues and defining what constitutes various kinds of attacks are reviewed, which are likely to be important if there is need for a multinational response. Because the most efficient application of radiation requires the total dose be well matched to the sensitivity of the microorganism(s) concerned, there were several valuable reports detailing progress on precise, accurate, rapid, and field-ready diagnostics to assay the type of microbial contamination. A strength of this conference was the inclusion of facility operators and experts on process control, safety, and dosimetry. Their operational knowledge, detailed information on the current state of the art, descriptions of facility capabilities, explanation of dosimetry standards, and presentation of available technology and emerging techniques provide a strong technical base. Only from such a technical base is it possible to consider what resources are available, determine those that could be used most effectively in any particular situation where there has been the illicit use of biological agents, and provide a high degree of assurance of the effectiveness of the decontamination effort. Also addressed was the radiation sensitivity of several types of agents of concern, including bacteria, bacterial spores, and viruses. Furthermore, factors that could alter an agent's radiation sensitivity were discussed. Several conference participants presented information on the U.S. response to the mail contamination, the approach that was taken, and some of the lessons learned. This conference also provided a forum for radiation experts on a broad regional basis to meet one another or become reacquainted. Potentially, this may be one of the most important facets of the conference. An important aspect of the U.S. response was rapidly making the needed connections and coordination among the appropriate scientists, private sector facility operators, and regulatory officials.

The conference recommendations were encapsulated in a formal memo to the International Atomic Energy Agency. In brief, the memo made following recommendations: (a) there is a need for a comprehensive assessment of the potential use of ionizing radiation for the destruction of biologically hazardous materials, (b) a need to assemble a committee of experts to develop and maintain a database on the use of radiation technology for biological agent defeat and to identify critical areas that still need to be addressed, (c) consider organizing an experts' meeting to advise the Coordinated Research Project on possible future Member States' actions, and (d) compile a list of radiation sources and locations capable of contributing to biological agent inactivation.

This workshop is a valuable basic reference for the use of radiation decontamination technologies against bioterrorism agents. The conference and its proceedings also provide a template for future highly cooperative and productive meetings to facilitate international interactions among those concerned with preparing responses to biological agent attacks. Hopefully these proceedings will stimulate support and foster collegial efforts in research on these technologies, which will not only improve their use in biological agent defeat but also broaden their applicability for medical and industrial processing.

Acknowledgements

The workshop was held at the National Center for Public Health, National Research Institute for Radiobiology and Radiohygiene, Budapest, Hungary for March 7 to 9, 2004. The workshop was attended by 43 participants from NATO, Partner and Mediterranean Dialogue Countries.

The editors would like to thank NATO Public Diplomacy Division for the support given both to organizing the conference and the publication of this book. We are especially grateful to Mr. F. Carvalho Rodrigues, Programme Director, Security-Related Civil Science and Technology, for his support. Without their financial assistance and support in working with us during the proposal and subsequent reviews, this workshop never would have taken place.

On behalf of the participants the editors also express their gratitude to the National Research Institute for Radiobiology and Radiohygiene and their staff for their superb effort in hosting this meeting. We would also extend our appreciation to Ms. Elisabeth Tóth for her editorial help.

L.G. Gazsó C.C. Ponta

Contents

Radiation Technology for New Materials Development, Human Health and Environment Protection

Andrzej G. CHMIELEWSKI and Mohammad HAJI-SAEID
Industrial Applications and Chemistry Section,
Division of Physical and Chemical Sciences,
Department of Nuclear Sciences and Applications,
International Atomic Energy Agency,
Wagramer Straße 5, 1400 Vienna, Austria

Abstract. Developments concerning radiation technologies' applications are discussed in the paper. The industrial irradiators based on gamma isotope and electron or X rays, accelerator driven sources are reviewed.

Present applications: polymers and rubber processing, sterilization and food irradiation are reported. Future possible developments in the field of natural polymers' processing and nanomaterials/nanomachines engineering are presented as well. The technological breakthrough achieved in the field of applications of radiation processes for environment protection illustrates new opportunities of the process utilization. This refers to the inactivation of biological warfare agents by ionizing radiation to the same extent. The economical and social aspects are shortly underlined.

Finally, the role of the Agency's programmes in promoting the above and the progress achieved is well described by this panoramic view of the status of radiation technology presented in this paper.

Introduction

Radiation has been discovered more than one hundred years ago. Since than, properties of radiation to modify physico-chemical properties of materials have found many applications. Radiation technologies applying gamma sources and electron accelerators for material processing are well established processes. There are over 160 gamma industrial irradiators and 1,300 electron industrial accelerators in operation worldwide. They are being widely used for sterilization, food irradiation and polymer processing. In the last 30 years, 648 industrial accelerators were installed in the U.S.A. and only 308 in Japan.

New developments in the field of radiation sources engineering are compact size gamma irradiators, high power electron accelerators (medium energy range) for environmental applications and other types (high energy range) for materials' processing, with direct e^-/X conversion. Future applications of low energy, inexpensive EB processing systems are foreseen. Electron beam lithography for microelectronics is a well-established technique.

The already tested e^-/X system equipped in an accelerator of 700 kW power opens new horizons for this kind of application. The developments described above need introduction of new computational methods that facilitate prediction of dose distribution, even in containers filled with complex products of varying densities. This technique provides good solutions for homeland security applications which may be complemented by mobile system applications.

Technologies to be developed besides environmental applications could be nanomaterials, structure engineered materials (sorbents, the composites, ordered polymers, etc.) and natural polymers' processing. New products based on radiation processed polysaccharides have already been commercialized in many countries of the East Asia and Pacific Region, especially in those being rich in natural polymers.

Very important and promising applications concern environment protection – radiation technology being a clean and environment friendly process, helps to curb pollutants' emission as well. Industrial plants for flue gas treatment have been constructed in Poland and China. The pilot plant in Bulgaria using this technology has just started its operation. The Polish plant is equipped with accelerators of over 1 MW power, a breakthrough in radiation technology application. The industrial plant for wastewater treatment is under development in Korea and a pilot plant for sewage sludge irradiation has been in operation in India for many years.

At the beginning of the 21st century, new science and technology development programmes are being elaborated and implemented, including UN resolutions concerning sustainable development, Johannesburg protocol, 6th EU Thematic Framework and others. Therefore, the Agency's Industrial Applications and Chemistry Section of the Division of Physical and Chemical Sciences, Department of Nuclear Sciences and Applications, organized a Technical Meeting (TM) on "Emerging Application of Radiation Technology for the 21st Century" at its Headquarters in Vienna, Austria, in April 2003, to review the present situation and possible developments of radiation technology to contribute sustainable development [1]. The meeting gathered the most eminent experts in the field and future programmes were discussed and recommendations elaborated. This meeting provided the basic input to launch others on the most important fields of radiation technology applications. The first one on "Advances in Radiation Chemistry of Polymers" was held in Notre Dame, U.S.A., in September 2003 [2], the second on "Status of Industrial Scale Radiation Treatment of Wastewater" in Taejon, Republic of Korea, in October 2003 and the third on "Radiation Processing of Polysaccharides" in Takasaki, Japan, in November 2003. The Agency actively participated in the IMRP 2003 held in Chicago, U.S.A., where future programmes in the field were discussed. Finally, the meeting on "Emerging Applications of Radiation in Nanotechnology" will be held in March 2004 in Bologna, Italy. The Agency hopes that the outcome of this meeting will initiate a new programme and international collaboration for research concerning application of various radiation techniques in the nanotechnology field. This should bridge radiation specialists with other research groups in the field and make connections between programmes of the Agency and big international and national projects.

The critical reassessment of R&D works and industrial applications of the technology are being reviewed in the paper.

1. Radiation Sources

The number of irradiators, working on the service basis or installed on-line is growing. The Agency has prepared a directory for industrial gamma irradiators [3] and plans to prepare a similar database for electron accelerators.

1.1. Gamma Irradiators

The number of irradiation units (approx. 160, the number was given in the speech of J. Mesfield during the opening of IMRP, Chicago, 2003) increased remarkably since the last report of Machi [4]. The world directory covering information on the 121 industrial and semi-industrial gamma irradiators prepared by Mehta [3] reports 17 new units commissioned in the years 2000-2002. The big irradiators with an activity over 1 MCi comprise over 20% of the total number. The other directory which has been prepared by Nordion is listing 64 plants [5].

1.2. Electron Accelerators

The total number of accelerators installed all over the world exceeds 13,000, among them the number of units applied for radiation processing being close to 1,200. Direct, transformer accelerators, single resonant cavity accelerators and microwave source powered linear accelerators have been found to be the most suitable for radiation processing [6]. The industrial accelerators' development is still in progress, not only due to new areas of application but also because of demands of lower cost and more compact size machines [7, 8]. Some new countries elaborated their own programmes concerning accelerator family developments. The low energy accelerators' capability has not been explored fully up to now [9].

New environmental applications demand development of high power, reliable accelerators. The most powerful radiation processing facility, applying accelerators over 1 MW total power, has been constructed for power plant emitted flue gases purification [10]. However, these new challenges for accelerator manufactures demonstrated that further progress in accelerator technology is needed and possible.

1.3. Electron Beam Units Equipped in e^-/X Converters

Application of X-rays for radiation processing based on X-ray tubes is quite popular in the case of blood irradiation. The concept of e^-/X conversion is known for years, a lot of R&D was performed in the field and some units were installed [11].

However, a breakthrough in technology is expected after implementation of the high power units which are already being tested (Jongen in [1]). Commercial irradiators are being offered on the market (Cleland in [2]).

2. Radiation Processing

Chemical or material engineering mostly apply high temperature and/or high pressure processes for material synthesis/modification and quite often a catalyst is required to speed up the reaction. Radiation is the unique source of energy which can initiate chemical reactions at any temperature, including ambient, under any pressure, in any phase (gas, liquid or solid), without the use of catalysts. However, the temperature rise factor should be considered when material is processed with the high dose [12].

2.1. Materials Modification

2.1.1. Synthetic Polymers

Among irradiated materials, polymers are the most prominent ones. Through modification or being a main component of, they can be found in radiation sterilized medical products. Therefore, changes in their structure may either be beneficial or undesirable. These facts are the reason why R&D concerning these materials is broad and most developments are foreseen in this area.

The application of radiation for modification of synthetic materials, mostly curing and crosslinking is a well-established technology [13]. For example, the total number of electron accelerators used for radiation crosslinking in Japan is 111. 59 for wires and cables, 25 for tires, 16 for plastic forms, and 11 for heat shrinkable tubes.

Radiation crosslinking of wires and cables was commercialized by Raychem in 1957, followed by Japan (1964), Poland (1974), United Kingdom, France, Germany, China, Republic of Korea and other countries [14]. Now, this is the most active and largest area of radiation processing. The radiation crosslinked wires and cables show excellent heat resistance (long-term thermal stability and short-term thermal stability) as well as abrasion resistance. The main markets for radiation crosslinking of wires and cables are actually those requiring performances of wires at accelerated temperatures. Thus, a very well established market is the production of wire harnesses (assemblies of wires) for electronic instruments such as computers and audio-video instruments. Harnesses for automobiles are also a big market for radiation crosslinking. Huge amounts of radiation crosslinked wires and cables are incorporated in an automobile to reduce the total weight and to increase overall efficiency of the automobile. Radiation crosslinked heat resistant wires and cables are an essential component in the engine room. Polyurethane covering the outside jacket of sensor cable for anti-lock brake system is also radiation crosslinked to improve the resistance against hot water. Other radiation crosslinked wires used in automobile are lead wires for high power lamp and non-halogen flame-retardant lead wires. Insulating materials used in wires and cables crosslinked by radiation include PE, PVC, fluoropolymers and polyurethane. PVC consumption tends to decrease as compared to polyolefin due to the recent concerns relating to the environment originating from the carcinogenic nature of some plasticizers and the toxic halogen fumes generated on incineration.

The polyethylene foam is used in a wide range of fields, such as automotive, buoyancy, flotation, insulation and packaging because it is easy to form and has superior heat insulation, flexibility and cushioning properties. In the foaming process, the melt viscosity suddenly decreases when it is heated above its melting point, in which case the gas generated from the foaming agent cannot be retained inside the resin. This makes it difficult to obtain foam with good expansion ratio, making it impossible to control the size and the number of cells. If crosslinking is used, more control can be exercised over the formation of cells with good control of nucleation and ultimate cell size. The manufacturing methods of crosslinking polyethylene foams are classified into two categories based on a type of crosslinking. One is chemical crosslinking by using peroxides as a crosslinking agent, and the other method is crosslinking by irradiation. Japan is a pioneer in commercialization of the latter technology.

This is crosslinked, closed-cell polyolefin foam developed by Sekisui Chemical Co. Ltd. Now several companies manufacture plastic foams by radiation crosslinking technique in Japan. Polyolefin foams are lightweight, heat insulating, shock absorbing, highly mouldable, and demonstrate non-water-absorbency. Due to these properties, polyolefin have a wide range of applications such as moulded interior automobile components, insulation materials for construction uses, joining materials, pipe covers, miscellaneous industrial materials, consumer goods, and healthcare and sports products. Predominantly used polymers include PE, ethylene-vinyl acetate copolymer and polypropylene.

Since radiation technology is, like all others, product-oriented, developments in other fields (polymer tubing for household) increase the demand for irradiation services (Zyball in [1]).

2.1.2. Rubber and Natural Latex

Various components of a tire include body ply, inner liner, tread, side wall and bead. These lead to the construction of green tires (raw tire), which are then being expanded into mould by inflation and vulcanized at high temperature and pressure. The displacing of cord fabric in body poly will give a fatal defect to the final products. This requirement prenecessitates the use of thicker plies in order to assure the adequate thickness of the final product. However, precise control of thickness is needed in the production of radial tires to reduce the weight of final tire. Thus, the radiation pre-vulcanized body ply will not decrease in thickness or be displaced during subsequent construction and vulcanization of the tire.

Goodyear and Firestone introduced this technology development in 1957, followed by Firestone who first started commercial application in the early 70s, Bridgestone was the first in Japan to install the first electron accelerator. Six companies produce about 170 million tires in Japan according to estimates of 1997. Five companies have installed electron accelerators for pre-vulcanization of carcass ply to increase green strength. Typical electron accelerator used in tire industry is low energy one like 500 keV, current 75 mA to 150 mA. Irradiation doses are 15-50 kGy. One accelerator can treat 30,000-50,000 piles/day.

In a conventional process, crosslinking or vulcanization is carried out by sulphur and heating. A small amount of the toxic substance nitrosoamine, formed during vulcanization, remains in the product. Radiation vulcanization leads to products with improved mechanical properties as compared to sulphur or peroxide crosslinking. Technology of radiation vulcanization of natural rubber (NR) latex (RVNRL) was developed at TRCRE by Makuchi at el [15]. The necessary dose for vulcanization is 15 kGy. The crosslinking sensitizer is n-butyl acrylate. The products are extremely safe due to the absence of N-nitrosamines. A low toxicity and smaller amount of extractible proteins are the merits of the technology. Pilot plants of RVNRL have been set up in Indonesia, India, Malaysia and Thailand for the production of dipped articles such as examination gloves, surgical gloves and balloons.

2.1.3. Natural Polymers

The success of radiation technology for the processing of synthetic polymers can be attributed to two reasons, namely the ease of processability in various shapes and sizes and secondly, most of these polymers undergo crosslinking reaction upon exposure to radiation. On the other hand, naturally occurring polymers were difficult to process and degraded when exposed to high energy radiation. Thus, the area of radiation processing of natural polymers largely remained unexplored and industrial applications have been difficult to achieve. A lot of research work especially related to the use of radiation technology for minimizing the environmental pollution associated with the processing of natural polymers such as dissolution of cellulose in the viscose-rayon process was carried out in countries like Canada and the Russian Federation [16]. However, the process could not be commercialized as the viscose producing plants were relocated to countries in the South-East Asia region where environmental issues were less demanding. In recent years, natural polymers are being looked at again with renewed interest because of their unique characteristics like inherent biocompatibility, biodegradability and easy availability. Traditionally, the commercial exploitation of natural polymers like carrageenans, alginates, starch, etc. has been based, to a large extent, on empirical knowledge. But now, the applications of natural polymers are being sought in knowledge-demanding areas such as pharmacy and biotechnology which are acting as a locomotive for further scientific research in their structure-function relationship. This opportunity was realized a few years ago by Member States of the RCA region and a project on utilization of naturally occurring polymers for value addition through radiation processing was initiated in the RCA region in 1998. The project is especially of relevance to the region keeping in view the available natural resources and the prevailing socio-economic conditions. A variety of natural polymers occurs in the region such as carrageenans, alginates, starch, etc. Besides this, it was also envisaged to utilize the large amount of agricultural waste (byproducts) generated in the region for value addition. Under the project, work was focused to achieve the following:

- Development of applications of low molecular weight polysaccharides in the area of agriculture as plant growth promoter.
- Utilization of specific properties of natural polymers as an additive to produce radiation processed hydrogels.
- Development of methods for producing radiation crosslinked completely biodegradable hydrogels.
- Exploration of possibility of using radiation modified materials or methods for environmental remediation.

As mentioned earlier, naturally occurring polymers or their derivatives undergo, under normal conditions of irradiation, a chain scission reaction. The work done at JAERI has shown that under suitable conditions, polymers such as carboxymethyl cellulose (CM-cellulose) or carboxymethyl starch can be crosslinked to form a hydrogel material [17]. This offers an opportunity to obtain non-toxic, additive free, totally biodegradable and biocompatible crosslinked hydrogels for many applications. This work has resulted in the development of a non-bedsore forming hydrogel mat based on radiation crosslinked CM-Cellulose hydrogel. The product was extensively tested in collaboration with a local hospital and found to be extremely useful. The technology for the product has been transferred to a private manufacturer and the product has been successfully commercialized. Similarly, radiation crosslinking of chitosan in aqueous solutions has been achieved by irradiating it in presence of a sensitizer, carbon tetrachloride. The crosslinked material is now being evaluated as an adsorbent for removing toxic materials from the waste waters. Formation of crosslinked polymer hydrogels from natural polymers and their derivatives offers new avenues for applications of such materials in many areas [18].

The healing of wounds, especially burn wounds has been a challenging medical problem as such wounds take a long time to heal and need to be protected to prevent infection. A radiation processed wound dressing based on PVP, agar and polyethylene glycol is well established on the market. Naturally occurring polymers like alginates and carrageenans are known to possess excellent wound healing characteristics. In order to utilize the functional properties of these polymers for wound healing, a PVA based hydrogel containing naturally occurring polymers like agar and carrageenan has been developed and commercialized in India. Similarly, a PVP based hydrogel containing carrageenan has been developed and extensively tested on patients in the Philippines with very encouraging results. An additional feature of these hydrogels is their usefulness in curing wounds of patients with diabetic ulcers which are otherwise difficult to treat. The PVP-carrageenan gel has the additional advantage of being a haemostatic agent which can be extremely helpful in many medical emergencies. The successful development of such materials provides new opportunities for the use of natural polymers in knowledge based technologies [19].

Low molecular weight naturally occurring polysaccharides like chitosan and alginates, prepared by conventional methods, have been reported to possess novel features such as promotion of germination and shoot elongation and stimulation of growth of Bifidiobacteria. As compared to conventional techniques like acid or base hydrolysis or enzymatic methods, radiation processing offers a clean one-step method for the formation of low molecular weight polysaccharides in aqueous solutions even at high concentrations. A lot of studies have been carried out in countries like Japan, Vietnam, China, India and the Philippines to investigate the plant growth promotion and plant protection effect of radiation processed polysaccharides in a variety of crops under different environmental conditions [20]. The results of these studies have clearly shown that radiation processed polysaccharides even at very low concentrations of few tens of ppm are very effective for use as plant growth promoter. This application offers tremendous opportunity to use them as highly effective organic fertilizers. Their biodegradability will be an additional advantage of using such materials as plant growth promoter. In Vietnam, three such formulations have already been commercialized namely, Olicide, Gold Rice (chitosan based) and TID (alginate based).

The antibacterial and antifungal effect of low molecular weight chitosan has been demonstrated for reducing the post harvest losses by prolonging the shelf life of many fruits and vegetables by coating them with these polysaccharides. More importantly, low molecular weight chitosan has been demonstrated to be an effective plant protector against infectious diseases and environmental stress [19].

2.1.4. Semiconductors

Precise control over carrier lifetime is an essential factor in meeting the ever-increasing market expectations for power semiconductor device performance. Diodes and thyristors, in all power categories, which did not have the required switching and release times after diffusion, can be properly adjusted by irradiation, thus, saving them from rejection [21]. Proper adjustment of the switching time in the case of high-power bipolar semiconductor devices [22] gives remarkable electricity savings during operation of controlled devices, e.g. electrical engines. Sometimes both, electron and ion beam treatments, are being used in a combination to optimize switching characteristics. Electron treatment allows control over carrier lifetime throughout the device whereas the much more limited penetration of ions permits the carrier lifetime to be precisely changed in a specific region within the device. Electron doses range from 0.05 to 400 kGy and ion (e.g. proton and helium) doses from 1E9 to 1E13 ions/cm^2.

2.1.5. Other Applications

Of the various irradiation agents available, gamma rays are the preferred choice because they produce excellent uniformity of coloration, do not consume electrical power nor produce localized heating or induce radioactivity. With the exception of diamond and the possible exception of some blue turning topaz, the nature of the source is immaterial; the alteration produced is just the same as long as there is sufficient energy supplied by the irradiation. Typically, only one to 10 eV are needed. Though causes of color are diverse, in most cases the change involves a color centre.

In the U.S.A., there are several irradiation contract facilities which convert colorless topaz into blue topaz using an EB processing system involving sometimes a combination of treatment in a reactor where some cooling period for radioactivity produced in gems is required. However, nature of all irradiations is proprietary. In semi-precious stones, transformations include conversion of beryls into yellow and green beryl, of quartz into yellow (citrine), dark brown (smoky), purple (amethyst), topaz into imperial or blue topaz [23], spodumene into yellow or green, tourmaline (colorless) into pink red, zircon into brown to reddish and pearl into valuable brown, blue or black.

It may be remarked that deep coloration of certain gems such as blue tourmaline and dark green emerald has been introduced with a judicious combination of heat and radiation.

2.1.6. Expected Future Developments

Radiation as a tool for product engineering, like sensors [24] or membranes [25] is a still not fully exhausted area of application.

A further progress concerning natural polymers processing is foreseen. Many processes of radiation treatment of natural polymers, though known for a long time, have not yet been commercialized either because of the high cost of irradiation (high dose) or because of the reluctance on part of the industry to adapt to the radiation technology. It is therefore of importance to consider combining the beneficial effects of conventional technology along with radiation technology to overcome such problems. The use of electron beam processing of cellulose pulp for reducing its degree of polymerization (DP) has been studied very extensively as it offers advantage in terms of reducing the concentration of carbon disulphide in the viscose process. The present results show that a combination of radiation technology with an enzymatic treatment can be used to dissolve cellulose pulp even without addition of the toxic carbon disulphide solvent used in the conventional viscose-rayon process. The developed process utilizes the electron beam treatment for initially reducing the DP to a desired level and enzymatic modification to produce alkali soluble cellulose. The combined process offers many unique advantages over the existing viscose-rayon process which is now facing stiff environmental regulations due to large pollution associated with this process. Similarly, the cost of production of radiation processed low molecular weight chitosan is still high because the dose required to produce oligo-chitosan is rather high. Chemical treatment combined with radiation treatment can be useful in producing low molecular weight chitosan in a more cost effective manner.

An example for products obtained by natural polymer radiation processing is sorbents utilized in wastewater treatment. The complex forming ability as well as interaction with the negatively charged surfaces of polymers like chitosan makes them ideal candidates for use as adsorbents and flocculants. Many of these applications are currently being explored. However, the limited solubility of chitosan in near neutral and basic solutions has been a limiting factor in flocculation applications. Radiation processing offers unique advantages in terms of providing low molecular weight chitosan having nearly the same charge and higher solubility. On the contrary, the solubility of chitosan films or fibres in acidic solutions is a hindrance in its application as an adsorbent for removing toxic metals and dyes from industrial effluents. In this case, radiation technology can provide crosslinked or grafted chitosan for use as adsorbents.

These materials are the raw substrates for manufacturing of biodegradable plastics and radiation can be one of the tools used in the technology [26].

Finally, nanotechnology is one of the fastest growing new areas in science and engineering. It is predicted to have a major impact on the manufacturing technology in 20 to 30 years from now. The subject arises from the convergence of electronics, physics, chemistry, biology and materials science to create new functional systems of nanoscale dimensions. Nanotechnology deals with science and technology associated with dimensions in the range of 0.1 to 100 nm. To achieve technological progress, firstly, the underpinning core science will need to be established. An interdisciplinary approach is required, bringing together key elements of biology, chemistry, engineering and physics. The development of appropriate interdisciplinary collaboration is expected to present challenges no less demanding than the science itself.

Therefore, such collaboration from the side of radiation chemists and physicists is needed as well. They are not newcomers in the field, arrangement of atoms and ions has been performed using ion or electron beams and radiation for many years. Talking about nanotechnology, we have in mind materials (including biological ones) and nanomachines. Molecular nanotechnology is perceived to be an inevitable development not to be achieved in the near future. In this context, self-assembly and self-organization are recognized as crucial methodologies. Radiation chemists in general, regarding materials processing, presented in the past a similar approach as did chemists, namely, treatment in bulk. However, new trends concerning a more precise treatment technology were observed as well: surface curing, ion track membranes and controlled release drug delivery systems are very good examples of such developments. The last two products from this list may even fit into the definition of nanomachine: they control substance transport rate by their own structure properties. The fabrication of nanostructures yields materials with new and improved properties; both approaches, "top-down" and "bottom-up" can be studied. The ability to fabricate structures with nanometric precision is of fundamental importance to any exploitation of nanotechnology. Nanofabrication involves various lithographies to write extremely small structures. Radiation-based technology using X-rays, e-beams and ion beams is the key to a variety of different approaches to micropattering, synchrotron radiation is one of the tools [27].

Radiation effect on resists occurs through bond breaking (positive resist) or crosslinking between polymer chains (negative resist). Polymer is becoming better or less soluble in developer. This technique has already been commercialized. Due to the small wavelength of the 30 – 100 keV electrons, the resolution of electron beam nanolithography is much higher than that of optical lithography. To improve resolution, electron direct writing systems applying electrons with the energy as low as 2 keV are proposed to reduce electron scattering effects. Other studies concern formation and synthesis of nanoparticles and nanocomposites. Radiation synthesis of copper, silver and other metals' nanoparticles is studied [28,29]. The solution of metal salts is exposed to gamma rays and formed reactive species reduce metal ion to zero valent state. Formation of aqueous bimetallic clusters by gamma and electron irradiation was studied as well. Metal and salt – polymer composites are synthesized by this method. Metal sulphide semiconductors of nanometric matrices are prepared using gamma irradiation of a suitable solution of monomer, sulphur and metal sources. These products find application in photoluminescent, photoelectric and non-linear optic materials. An interesting field of radiation nanotechnological application concerns the development of PC-controlled biochips for programmed release systems. Nano-ordered hydrogels based on natural polymers as polysaccharides (hyaluronic acid, agrose, starch, chitosan) and proteins (keratin, soy-bean) being pH and electric potential responsive materials for such biochips and sensors. To avoid regress in the further advance of radiation processing of natural polymers, the nano approach to these biological materials should be developed further. Their self-organization and functionalism depend on the basic fundamentals of the discussed science. The studies on natural rubber-clay composites and thermoplastic natural rubber-clay composites have given interesting results. Nanomaterials with high abrasion and high scratch resistance will find industrial application.

2.2. Sterilization

Radiation sterilization is a well established technique [30] and most of its strong and weak points were addressed [31]. The observed tendency concerns multi-technique offer (radiation, EtO, heat) of the service. Gamma source [32] or electron beam [33] based service plants are operated in many countries. Process control and dosimetry are essential in these applications [34].

2.3. Food Irradiation

The acceptance of the use of radiation processing for the treatment of the food varies throughout the world. In the U.S.A., there seems to be greater public acceptance of food irradiation and related industry support. However, other regions, such as the European Union [35], seem reluctant to adopt some well-accepted practices of radiation treatment of certain food, such as spices, even though elsewhere, e.g. in North America, such irradiated products are in common use [36].

An accelerator equipped plant for food irradiation, processing mostly spices, is operated in Poland [37]. Decontamination of other products like meat lyophilised products is investigated as well [38]. The implementation of food irradiation must be accompanied by the development of a food detection laboratory [39].

3. Environmental Applications

3.1. Water, Wastewater and Sludge

Many R&D works have been performed concerning application of radiation technology for drinking water and wastewater purification. The plant for liquid sludge hygienization, furnished in cobalt-60 gamma source, is in operation in India [40]. The accelerator can be used for dewatered sludge treatment as well [41].

The pilot plant for dye factory wastewater treatment equipped in an electron accelerator has been constructed in South Korea [42] and an industrial project aiming at the treatment of 10,000 cubic meters of effluent per day is in progress.

3.2. Flue Gases

The electron beam technology for flue gas treatment has been developed by Machi and Tokunaga (Machi in [1]) in the early 80s. Lateron, this process was investigated in pilot scale in the U.S.A., Germany, Japan and Poland. Research on the pilot plant in Kaweczyn (20,000 cubic meters of flue gas per hour, two accelerators; 50 kW, 700keV each) was carried out. IAEA-JAER-NEK has constructed another pilot plant (flow 10,000 cubic meters per hour; 3x30 kW, 800keV) in Bulgaria to treat high humidity, high SOx gases from combustion of low grade lignite. The test operation has started.

The electron beam flue gas treatment plants are operating in the coal-fired plants in China and Poland [43]. Since the power of accelerators installed at the Polish plant is bigger than 1 MW, it is the largest irradiation facility ever built. The high efficiency of SOx and NOx removal was achieved and by-product is a high quality fertilizer. The other possible application of the technology is VOC and PAH treatment, e.g. in municipal waste incinerator plants flue gas purification units [44, 45].

The advantage of this technology over conventional ones has been clearly demonstrated from both, the technical and the economical points of view. Further, its implementation depends on the development in construction of reliable, high power accelerator with short maintenance time requirements.

4. Inactivation of Biological Warfare Agents

Ionizing radiation has been known to be very effective in the decomposition or inactivation of pathogenic microorganisms and protozoan parasites as such or in combination with other agents, such as ozone, heat, etc. Literature data available on this subject indicate that a much higher dose of radiation is required to kill viruses and bacterial spores than to kill pathogenic bacterial cells. Certain environmental factors are also able to influence the actual radiation response. The responses of cells to a given dose can be altered in different ways. This is possible because response depends on physical factors (quality of radiation, temperature, etc.) on chemical factors (oxygen, water content, chemical agents, etc.) and the biological or physiological factors (growth phase, amount of DNA). Spore forming micro-organisms have proven to be the most radiation resistant, particularly their spores, then in descending order of radiation resistance, yeast, mould, viruses and bacteria. Some highly resistant spores show D_{10} values as high as 10 kGy which requires a dose of 40 kGy to achieve a reduction of 4 log. This may be the case with some biological agents. In view of the recent developments in electron beam technology, and with proper design principles and under-beam handling systems, the use of electron accelerators seems to be capable of providing a good solution to decontaminate biological warfare agents. It is also gratifying to note that there are no significant engineering problems to build a transportable e-beam system for on-site decontamination works.

The Agency poses technical and organizational expertise to implement or coordinate the programme concerning the process. The studies on the subject have already been performed by the Agency.

5. Economical and Social Aspects of Radiation Technology Application

The good presentation of economic scale of utilization of radiation technology in the two biggest world economies; the U.S.A. and Japan is given in [46]. The gross values of medical products sterilized by radiation are equal to 4.8 b$ in the U.S.A. and 2.3 b$ in Japan, radiation curried radial tires values are 13.5 b$ and 8.4 b$ respectively. The market of semiconductors processed with application of this technology is equal to 37.2 b$ in the U.S.A. and 28.4 b$ in Japan.

Sterilization, disinfection, environmental applications and clean radiation (in comparison to heat and chemical processes) are all benefits of radiation applications for mankind and the environment [47].

6. Conclusions

Taking into account the total value of products in which manufacturing or modification radiation is used, the economic scale of radiation technology application is enormous. Some well established technologies like sterilization, polymers and semiconductor modification are in common use, often on-line.

High power accelerators, including those equipped in e^-/X converters, low energy systems and new compact gamma irradiators have been developed.

The new environmental applications of high power accelerators for gaseous and liquid effluents have been reported. All this experience matured this technology to a status at which it can be used for human hostile biological agents deactivation as well. Remote control or mobile units can be applied.

New developments follow the general trends of technological development and social needs, like nanotechnology, natural and structural polymers and environment protection.

All discussed examples illustrate the big role of radiation technology in industrial development, health and environment protection.

References

[1] IAEA, 2004, Emerging applications of radiation processing. TECDOC-1386, Vienna, Austria.

[2] IAEA, 2003, Report from a technical meeting; Advances in radiation chemistry of polymers. Notre Dame, Indiana, USA.

[3] IAEA, 2003, Directory of Commercial Radiation Processing Facilities in Member States, DGPF/CD Vienna, Austria.

[4] Machi, S., 1995, Radiation technology for sustainable development. Rad. Phys. Chem. 46(4-6), 399-410.

[5] NORDION, 2003, Supplies of Contract Irradiation Services. Ottawa, Canada.

[6] Zimek, Z., Chmielewski, A.G, 1993a, Present tendencies in construction of industrial accelerators applied in radiation processing. Nukleonika., 38(2), 3-20.

[7] Zimek, Z., Rzewuski, H., Migdal, W., 1995, Electron accelerators installed at the Institute of Nuclear Chemistry and Technology. Nukleonika., 40(3), 93-114.

[8] Cleland, M.R., Parks, L.A., 2003a, Medium and high-energy electron beam radiation processing equipment for commercial applications. Nucl. Instrum. Methods in Phys. Res. Sec. B, 208, 74-89.
[9] Berejka, A.J., 2003, Advances in self-shielded accelerators. In: IAEA, 2003a, 78-86.
[10] Chmielewski, A.G., Tyminski, B., Zimek, Z., Licki, J., 2001, Flue gas treatment by electron beam technology. Modern Power Systems. May, 53- 54.
[11] Migdal, W., Malec-Czechowska, K., Owczarczyk, B., (1996), Study on application of e^{-}/X convertor for radiation processing. Nukleonika, 4(1), 57 – 76.
[12] Cleland, M.R., Parks, L.A., Cheng, S., 2003b, Application of accelerators for radiation processing of materials. Nucl. Instr. and Meth. in Phys.Res.Sec.B, 208, 66-73.
[13] Drobny, J.G., 2003, Radiation Technology for Polymers. CRC Press, New York.
[14] Chapiro, A., 2002, Polymer irradiation: past-present and future. Rad. Phys. Chem. 63(3-6), 207-209.
[15] Makuuchi, K., 2003, An introduction to radiation vulcanisation of natural rubber latex. T.R.I.Global Co., Ltd, Bangkok.
[16] Iller, E., Kukielka, A., Stupinska, H., Mikolajczyk, W., 2002, Electron beam stimulation of the reactivity of cellulose pulps for production of derivatives. Rad. Phys. Chem., 63(3-6), 253-257.
[17] Fei, B., Wach, R.A., Mitomo, H., Yoshii, F., Kume, T., 2000, Hydrogel of biodegradable cellulose derivatives. I. Radiation-induced crosslinking of CMC. J. Appl. Pol. Sc. 78, 278-283.
[18] Zhao, L., Mitomoto, H., Nagasawa, N., Yoshii, F., Kume, T., 2003, Radiation synthesis and characteristic of the hydrogels based on carboxymethylated chitin derivatives. Carbohydrate Polym., 51, 169-175.
[19] Kume, T., Nagasawa, N., Yoshii, F., 2002, Utilization of carbohydrates by radiation processing. Rad. Phys. Chem., 63, 625-627.
[20] Hien, N.Q., Nagasawa, N., Tham, L.X., and al., 2000, Growth-promotion of plants with depolymerised alginates by irradiation. Rad. Phys. Chem., 59, 97-101.
[21] Fuochi, P.G., 1994, Irradiation of power semiconductor devices by high energy electrons: The Italian experience. Rad. Phys. Chem., 44(4), 431 – 440.
[22] Mittendorfer, J., Zwanziger, P., 2000, Application of statistical methods (SPC) for an optimised control of the irradiation process of high-power semiconductors. Rad. Phys. Chem., 57(3-6), 629-634.
[23] Ying, W., Yong-bao, G., 2002, Research on radiation induced color change of topaz. Rad. Phys. Chem. 63(3-6), 223-225.
[24] Trakhtenberg, L.I., Gerasimov, G.N., Aleksandrova, L.N., Potapov, V.K., Photo and radiation cryochemical synthesis of metal-polymer films: structure, sensor and catalytic properties. Rad. Phys. Chem., 2002, 65(4-5), 479-485.
[25] Mazzei, R.O., Smolko, E., Torres, A., Tadey, D., Rocco, C., Gizzi, L., Strangis, S., 2002, Radiation grafting studies of acrylic acid onto cellulose triacetate membranes. Rad. Phys. Chem., 64(2), 149-160.
[26] Zhai M., Yoshii F., Kume, T., 2003, Radiation modification of starch-based plastic sheets. Carbohydrate Polym., 52, 311-317.
[27] Hirota, Y., 2003, LIGA process – micromachining technique using synchrotron radiation lithography – and some industrial applications. Nucl. Instr. Meth. Phys. Res., 208, 21 – 26.
[28] Ila, D., Williams, E.K., Zimmerman, R.L., at al., 2000, Radiation induced nucleation of nanoparticles in silica. Nucl.Instr.Meth.Phys.Res.Sec.B.,166-167, 845-850.
[29] Joshi , S.S., Patil, S.F., Iyer, V., Mahumuni, S., 1998, Nanostr.Mat., Radiation induced synthesis and characterization of copper nanoparticle. 10(7), 1135-1144.
[30] Fairand, B.P., 2002, Radiation Sterilization for Health Care Products – X-Ray, Gamma and Electron Beam. CRC Press, New York.
[31] Morrissey, R.F., Herring, C.M., 2002, Radiation sterilization: past, present and future. Rad. Phys. Chem. 63 (3-6), 217-221.
[32] Katusin-Razem, B., Mihaljevic, B., Razem D., 2003, Microbiological decontamination of cosmetic raw materials and personal care products by irradiation. Rad. Phys. Chem., 66, 309-316.
[33] Zimek, Z., Walis, L., Chmielewski, A.G., 1993b, EB industrial facility for radiation sterilization of medical devices. Rad. Phys .Che., 42(1-3), 571-572.
[34] Zimek, Z., Kaluska, I., 2002, Sterilization dose auditing for various types of medical products. Rad. Phys. Chem., 63(3-6), 673 – 674.
[35] Ehlerman, D.A.E., 2002, Current situation of food irradiation in The EU and forthcoming harmonization. Rad. Phys. Chem., 63(3-6), 277-279.
[36] Morehouse, K.M., Food irradiation – US regulatory considerations. Rad. Phys. Chem., 2002. 63(3-6), 281-284.
[37] Migdal, W., Walis, L., Chmielewski, A.G., 1993, The pilot plant for electron beam food processing. Rad. Phys. Chem., 42(1-3), 567-570.
[38] Migdal, W., Owczarczyk, B., Radiation decontamination of meat lyophylized products. Rad. Phys. Chem., 63(3-6), 371-373.
[39] Stachowicz, W., Malec-Czechowska, K., Dancewicz, Z., Szot, Z., Chmielewski, A.G., (2002), Accredited laboratory for detection of irradiated foods in Poland. Rad. Phys. Chem., 63(3-6), 427-429.

[40] IAEA, 2002, Irradiated sewage sludge for application to cropland, TECDOC-1317, Vienna, 238 pp.
[41] Chmielewski, A.G., Zimek, Z., Bryl-Sandalewska, T., et al., 1995, Disinfection of municipal sewage sludges in installation equipped with electron accelerator. Rad. Phys. Chem., 46(4-6), 1071-1074.
[42] Han, B., Ko, J., Kim, J., at al., 2002, Combined electron beam and biological treatment of dyeing complex wastewater. Rad. Phys. Chem. 64, 53-60.
[43] Chmielewski, A.G., Iller, E., Tyminski, B., Zimek, Z., Licki, J., 2001, Flue gas treatment by electron beam technology. Modern Power Syst. May, 53-54.
[44] Hirota, K., Hakoda, T.,Taguchi, M., at all., 2003, Application of electron beam for the reduction of PCDD/F emission from municipal solid waste incinerators. Environ. Sci. Techn., 37, 3164-3170.
[45] Chmielewski, A.G., Ostapczuk, A., Zimek, Z., Licki, J., Kubica, K., 2002, Reduction of VOCs in flue gas from coal combustion by electron beam treatment. Rad. Phys. Chem., 63(3-6), 653-655.
[46] Tagawa, S., Kashiwagi, M., Kamada, T., et al., 2002, Economic scale of utilization of radiation(I): Industry.J.Nucl.Sc.Techn.,39(9), 1002-1007.
[47] Chmielewski A.G, Haji-Saeid, M, 2004, Radiation technologies: past, present and future, Rad. Phys. Chem. 71(1-2), 2004.

Radiation Safety Principles and Requirements at Gamma- and Electron Irradiation Facilities

András KOVÁCS
Institute of Isotopes, Chemical Research Center,
Hungarian Academy of Sciences,
H-1525 Budapest, P.O.Box 77, Hungary

Abstract. Radiation safety of gamma- and electron irradiation facilities is of basic significance concerning the safety of personnel involved in construction, operation and maintenance, the safety of the products treated as well as the environment of the facility. The safety principles with respect to the design and construction of the facilities, requirements concerning the basic parts and the operation of the plants, the reliable control of running irradiation facilities, the role of emergency planning and the role of licences are discussed. Recent measures to increase safety are also mentioned.

Introduction

Radiation processing has become a worldwide accepted industry during the past few decades. The most widespread applications involve radiation sterilization of medical products and pharmaceuticals, food irradiation, polymer modification, environmental and biomedical applications. These technologies utilize gamma-, electron- or X-ray radiation resulting in nowadays about 200 high activity gamma irradiation facilities and about 1200 electron accelerators (0.1 – 10 MeV) in operation worldwide.

Due to the nature of ionizing radiation used in radiation processing safety considerations have got basic importance with respect to licensing, design, construction, operation and maintenance of irradiation facilities as well as to transport, loading and unloading of radiation sources. These safety requirements involve precautions to protect the personnel and customers of irradiation facilities as well as the inhabitants living in the neighbourhood of the facilities. These precautions are inevitable, since severe radiation exposure may result from loss of control of the radiation source. Safety measures are needed to protect the products to be treated and to ensure safe operation by protecting the basic parts of the irradiation facilities including control and auxiliary instrumentation and equipments. It is of basic importance since e.g. damage to the radiation source can lead to dangerous contamination.

The responsibilities of the operating organization as well as that of all relevant parties taking part in the design, construction, import and installation to establish an irradiation facility, have to be clearly identified and documented [1]. Special programmes (i.e. regulatory programme and an emergency response planning) must be prepared and documented.

This paper in general follows the structure of the Safety Series No. 107 on "Radiation Safety of Gamma and Electron Irradiation Facilities" edited by the International Atomic Energy Agency [2].

1. Irradiation Facilities

1.1. Gamma Irradiation Facilities

The activity of the radioactive source in an irradiator ranges from a few terabecquerels (1 TBq = 27 Ci) to more than 100 PBq (3 MCi). With respect to the design of irradiation facilities four general categories are defined as follows:

Category 1.: An irradiator in which the sealed source is completely enclosed in a dry storage container constructed of solid materials and shielded at all times. At such "container type" irradiation facilities human access to the sealed source and to the volumes undergoing irradiation is not physically possible in the designed configuration. Such irradiation facilities are usually used e.g. in research laboratories to treat small volumes of products.

Category 2.: A controlled human access irradiator in which the sealed source is enclosed in a dry container constructed of solid materials. The sealed radiation source is fully shielded when not in use and it is exposed within a radiation volume that is maintained inaccessible during operation by an entry control system. Such an irradiation facility is usually located within a concrete biological shield providing the necessary radiation protection during operation.

Category 3.: An irradiator in which the sealed source is contained in a water filled storage pool and is shielded at all times. Human access to the sealed source and the volume undergoing irradiation is physically restricted in the designed configuration and proper mode of use. Such design is usually employed in a pool irradiator in which the product to be irradiated is lowered in a water tight container to the bottom of the pool among or next to the source.

Category 4.: A controlled human access irradiator in which the sealed source is usually contained in a water filled storage pool (or rarely in a dry storage place shielded by suitable shielding materials, e.g. lead). The sealed source is fully shielded when not in use and the sealed source is exposed within a radiation volume that is maintained inaccessible during use by an entry control system. The irradiation volume is typically contained within a concrete biological shield providing the radiation protection during operation or exposure of the radiation source. This type of irradiation facility can be either batch or continuous type and the latter one is equipped with a product transport system to take the products in and out of the irradiation room without lowering the source into its storage position.

Beside these general categories the irradiation facilities can also be categorized according to the

- storage type of the radiation source (dry or wet),
- radiation volume (closed or panoramic),
- product transport system (tote, carrier, pallet, bulk).

1.2. Electron Beam Facilities

With respect to induced radioactivity, electron accelerators of energies equal or less than 10 MeV are considered in the present discussion. The main differences concerning the different types of electron accelerators are due to the mode of accelerating the electron beam and to the method of producing the necessary high voltages. For the present purpose the electron irradiation facilities are categorized as follows:

Category 1.: An integrally shielded electron beam unit with interlocks, where human access during operation is not possible due to the configuration of the shielding. These facilities operate usually in the electron energy range of 0.1 - 0.5 MeV, although there are a few examples of such facilities operating with much higher energies due to recent designs.

Category 2.: An electron beam unit located in a shielded room that are maintained inaccessible during operation by an entry control system. These facilities operate usually with electron energies higher than 0.5 MeV.

2. Safety Design Principles of Irradiation Facilities

2.1. Radiation Safety Objectives

In the design, installation, commissioning, operation, maintenance and decommissioning of an irradiation facility the following radiation safety objectives have to be taken into account:

- During normal operation, maintenance and decommissioning as well as in emergency situations radiation protection has to be optimized ensuring that the radiation exposure of both workers and the public is kept as low as reasonably achievable, economic and social factors are taken into consideration (ALARA principle).
- It is necessary to ensure that during normal operation, maintenance and decommissioning as well as in emergency situations the radiation exposure of both workers and the public is kept below the relevant dose limits as given in the Basic Safety Standards for Radiation Protection [3].
- It has to be ensured that the probability of events giving rise to significant exposures as well as the magnitude of such exposures are kept as low as reasonably achievable, economic and social factors are also taken into account.

2.2. Safety Philosophy

Since the design of an irradiation facility depends on its planned operation as well as on its category [4, 5] it is not possible to suggest specific design of the required degree of safety, but to present several design principles to be used - if necessary in combination - to achieve and maintain the required reliability. These basic design principles are the followings:

2.2.1. Defence in Depth Concept

This concept should be applied to all (organizational, design related, behavioural) safety activities to ensure that they are covered by overlapping provisions so that if a failure should occur, it would be compensated for or corrected.

The design procedure shall incorporate this concept so that multiple levels of protection have to be provided and the necessity of human intervention has to be minimized. In the course of the design process e.g. a series of levels of defence (in terms of equipment and procedures) is provided to prevent accidents or to mitigate their consequences in the event that preventive measures fail. The first level of defence is to prevent deviation from the normal operation to be achieved by soundly and cautiously designed, constructed and operated facility, by establishing and maintaining suitable quality assurance programme. The second level of defence is to detect and respond to deviations from normal operating conditions to prevent anticipated operational occurrences from escalating into accident conditions. The aim of the third level of defence is to mitigate the consequences of an accident and through achieving stable and acceptable conditions.

Irradiation facilities should be operated only in that case, if all levels of defence are in place and functioning.

2.2.2. Redundancy

It is the use of more than the minimum number of items needed to accomplish a given safety function. It enables the failure or unavailability of one item to be tolerated without loss of the function. A redundant component can also be considered as such an item, which is not really needed, but built in case of failure of another component of similar purpose.

2.2.3. Diversity

The reliability of some systems can be enhanced by using the principle diversity. It is applied to redundant systems or components that perform the same safety functions by incorporating different attributes (like different principles of operation, different operating conditions, etc.) into the systems or components.

2.2.4. Independence

Independence is achieved in the design of systems by applying functional isolation and physical separation. The reliability of redundant systems can be improved by using independent components or systems thus avoiding the failure or loss of other components, safety function or equipment designed to mitigate the effect of incidents in the case of the failure of one component or system. The use of this concept improves the reliability of systems also by maintaining independence of systems or components of different importance to safety.

2.2.5. Programmable Electronic Systems

Programmable electronic systems [6] are in use in safety control applications with potential problems related to the integrity of the hardware and validation of the software leading to faults in the system. Any alterations to the software can be made only in case of having authorization from a competent authority.

2.2.6. Safety Analysis

A formal method of safety assessment – e.g. a hazard analysis technique like a probabilistic safety analysis – should be used. Each component within the system should be considered in turn. The potential types of failure and their consequences for the system as a whole should be taken into account considering the reliability of the safety dependent operating procedures and should encompass both inadvertent and deliberate failure to follow procedures. The operating organization has to demonstrate to the competent authority how the irradiator design and the related operational procedures will contribute to the prevention and mitigation of accidents. This information has to be provided in the form of a documented safety analysis as part of the licencing document submitted to the regular authority.

3. Safety Design Requirements

With respect to the design of irradiation facilities guidance is needed for the designers in terms of dose or dose rate objectives since these parameters are in relation with the planned use of the facility. The permissible radiation levels outside the biological shield to which employees, customers or the public may be exposed during operation of the facility has to be based on the ALARA principle taking into account of any additional dose constraints that may have been specified for the purpose by the local competent authority. Experiences with respect to the operation and maintenance of similar facilities should be collected and used in specifying the levels of exposure that are achievable in practice. Such experience originating from many countries (Member States) has shown that irradiation facilities can be designed and operated in such a way, that radiation workers are exposed to radiation levels significantly less than 5 mSv per year. The basic design parameters and related safety features concerning the main parts of irradiation facilities (radiation source, biological shield, product transfer system, control and safety systems, auxiliary systems) are discussed briefly below.

3.1. Radiation Source

3.1.1. Design of Sealed Sources

The radiation source in gamma irradiation facilities in most cases is ^{60}Co isotope, while in a few cases it is ^{137}Cs isotope. These sources are double encapsulated sealed sources and the general requirements for such sources are given in ISO Standard 2919 [7, 8]. In addition to this standard the manufacturer and the user have to take into account the possible effects of fire, explosion, corrosion, and other aspects related to the continuous use of sealed sources (e.g. leak tests). Specific requirements are neded for radiation sources kept in water (wet storage conditions). Both the source manufacturer and the user have to maintain suitable certification and documentation of the sealed sources for their own purpose (record keeping) as well as for the competent authorities for licensing.

3.1.2. Internal Design

The design of the source holder and source rack has to ensure the fixed positioning of the radiation sources to avoid their dislodging. The source must be provided with suitable mechanical protection to avoid damage or interference by the product boxes or product carriers. The product positioning system shall be provided with controls to detect any malfunction of the system resulting in the source to automatically become fully shielded.

3.2. Biological Shielding

Direct radiation exposure from the operation of irradiation facilities shall be limited by appropriate shielding using materials of well established characteristics [5, 9]. Biological shield provides protection outside of the irradiation room (radiation volume) during operation and provides safe storage for the radiation sources. The amount of shielding should be determined by reference to any dose rate requirements specified by the competent authority. Shielding calculations for design purposes shall be carried out by specialists. In the case of wet storage irradiation facilities special care should be taken to pool integrity.

3.3. Control, Safety and Auxiliary Systems

These systems, devices and instruments are designed and used to control the safe operation of the irradiation facility (e.g. controlling the safe operation of the speed of the product transport system, position of product boxes, movement and position of the source hoist), as well as to control the safety systems (like dose level, water level and condition in the pool, ventilation, power supplies, etc.). These safety features are discussed according to the categorization of the IAEA Safety Series No. 107 [2].

3.3.1. Access to the Radiation Source and Interlocking System

Special attention – and suitable interlocking system - is needed to avoid personnel access into the irradiation room while the radiation source is in exposed position or the source hoist mechanism is energised. Interlocking systems and monitors are required at the personnel access door, at the product entry/exit ports as well as at the removable radiation room shield plugs.

Fixed radiation monitor (with visible and audible alarm signals) shall be provided to detect the radiation level in the radiation room. Interlocks shall be provided to control the source position and exposure system. In case of any malfunction access into the radiation room shall be prevented.

At wet storage facilities fixed monitoring systems (with audible alarm) are provided to control the columns of the water treatment system and to detect contamination originating from source leakage. The monitor shall be interlocked with the irradiation controls to ensure that the source returns to its shielded position, the water circulation is stopped and warning signal is given.

In case of fully shielded facilities (categories I and III for gamma and I for electron beam units) the irradiator shall not be operable until all shielding is in place and all other safety devices are actuated.

3.3.2. Control Console

Each irradiator shall have a master control (key operated switch, padlock, etc.) for use to prevent unauthorized operation. In addition means shall be provided to terminate an irradiation and return the irradiator to its safe ("source not in use") status at any time, and an emergency stop device should be available at the control console.

3.3.3. Radiation Room

The radiation room shall be equipped with a safety delay timer to automatically generate visible and audible signals to alert persons that the source exposure has begun. To protect persons inadvertently shut inside the radiation room, an emergency exit or a low dose location should be provided. An emergency stop device should also be located inside the irradiation room to terminate radiation or to prevent the source from moving from storage to exposed position.

3.3.4. Wet Storage Irradiators

In these irradiation facilities the integrity of the pool and the use of corrosion resistant pool components have to be ensured. Water level control with visible and audible signals has to be provided for such cases when the water level falls to more than 30 cm below the normal level. Furthermore automatic water replenishment, water cooling, water conditioning system with conductivity control, external high flow rate water supply have to be provided. To protect personnel a physical barrier has to be built along the pool.

3.3.5. Geological Site Considerations

Geological features that could adversely affect the integrity of the radiation shielding have to be evaluated considering the physical properties of materials under the irradiator site and its environs. The irradiator has to be equipped with seismic detector in the case of possible severe seismic activity.

3.3.6. Fire Protection

Heat build-up originating from anomalous operation of the facility can lead to combustion, which makes necessary the use of heat and smoke sensing devices with visible and audible alarms. The triggering of these devices shall cause the source automatically to become fully shielded and the product positioning and the ventilation system to shut down. A fire extinguishing system – not containing chemicals or corrosive materials – should be provided in the irradiation room.

3.3.7. Power Failure

Means shall be provided that in case of electrical or non-electrical (e.g. pneumatic or hydraulic) power failure the source shall automatically be returned to the fully shielded position and the irradiator shut down. The safety control system has to be designed so, that it will remain in operation in the case of power failure.

3.3.8. Ventilation

Ventilation system – which creates a negative pressure in the irradiation room - shall be used to protect personnel against exposure to concentrations of harmful gases (ozone, nitrous gases and other noxious gases) - above the prescribed threshold limit values – produced upon irradiation. This type of forced air system – with continuous air flow monitoring - should be interlocked so, that failure of the system will automatically shut the plant down together with the source moving to the fully shielded position. A time delay interlock mechanism should be introduced to prevent personnel from entering the irradiation room immediately after irradiation.

3.3.9. Warning Signs and Symbols

Clearly visible signs bearing the radiation symbol and warnings must be located at the personnel access door leading into the irradiation room. Clearly visible irradiation status indicators shall be provided at the control console to indicate the status of the source and these shall be visible at each personnel or product entry/exit location. Audible signals designed into the irradiator control system shall be distinct and loud enough to gain immediate attention of persons in the area.

Category I gamma irradiation facilities shall have clearly visible labels showing the contained radionuclides, their activities and the related dates.

4. Special Safety Requirements for Electron Accelerator Facilities

4.1. Safety Design Considerations

An objective of manufacturers of industrial accelerators is to design EB sources for simplicity and reliability of operation. With respect to potential radiation accidents care should be taken due to the possibility of generating X rays from dark currents, if the acceleration capabilities of the facility or its subsystems remain because of improper or partially disabled accelerating stages. To reduce the chance of hazardous occurrences the following features should be considered in the course of design:

- Positive means of disabling the main acceleration system;
- Built-in machine parameter monitoring;
- Built-in remote machine diagnostics

4.2. Shielding

Due to the differences in the nature of electron radiation and X-ray radiation generated only the latter should be considered when carrying out shielding calculations. These calculations must be performed for the maximum energy and current that the accelerator can produce. To minimize the generation of X-ray radiation low atomic number materials should be used as far as possible. Except for self-shielded accelerators and special purpose machines operating at higher energies ordinary concrete is preferred as shielding material. To calculate the shielding thickness different mathematical methods and special programmes are available [5, 10].

4.3. Other Requirements

The operating parameters of accelerators (voltage and current) should be interlocked with the product transport system. Testing and commissioning of the accelerators should be carried out at maximum operating parameters and with product transport system under the beam as close as possible to actual operating conditions.

5. Regulatory Control

5.1. Regulatory Programme

In the use of ionizing radiation regulatory programme – enforced by the local authority - is an important part of radiation protection [11]. This programme is intended to control irradiation facilities by systems of notification, registration or licensing, dependent on the legislation in any given country. The most common system of regulatory control is that of licensing or approval as described briefly here.

It is important to note that the main responsibility for safety belongs to the person(s) carrying out the given tasks (design, installation, operation, maintenance, decommissioning).

It is also essential that a regulatory system must be in place in any country before approval of an application to build an irradiation facility.

5.2. The Approval Process

The control of radiation safety concerning the siting, design, construction, commissioning, operation, maintenance and decommissioning of an irradiation facility is carried out by governmental licences or approvals, which authorize actions and place conditions on the applicant. The local competent authority has to review and assess the application to decide whether or not the approval is granted.

5.2.1. Functions of Approvals

This official document authorizes a certain activity(ies) in connection with all procedures relevant to the establishment and use of an irradiation facility and establishes requirements how to perform these activities.

5.2.2. Stages of Approval Process

The main stages of this process has to include the regulation of siting, design, construction, commissioning, operation, maintenance and decommissioning. Before approval is issued a radiation safety analysis (case) – as part of the overall licensing document - must be prepared by the applicant including the complete description of the facility with details of all important factors affecting its operation and safety.

5.2.3. Requirements for the Applicant

The applicant has the primary responsibility to ensure safety for all stages of establishing and operating an irradiation facility. The applicant has to demonstrate to the competent authority that workers, customers and public will be adequately protected. The manufacturer of the facility will also provide all necessary information (e.g. training of workers to operate the facility, removal of spent radioactive sources, construction and installation corresponds to the specifications, etc.) for the applicant. Other requirements involve the appointment of radiation protection personnel, assessment of hazards, periodic testing and surveys of the radiation protection and safety aspects of the facility, regular reports on senior staff changes, radiological data like personal dosimetry, contamination monitoring, etc.

5.2.4. Responsibilities of the Competent Authority

Based on the applicant's technical submissions the competent authority has to determine whether the design and the operating procedures comply with the actual safety objectives and requirements and whether the facility can be established and operated.

5.2.5. Review and Assessment of the Application

The basis for the review and assessment of the radiation safety implications of the proposed facility are those documents, safety reports, etc., which have been submitted by the applicant.

5.2.6. Programme of Review and Assessment

The competent authority has to prepare a programme of review and assessment which comply with the stages of the approval process. It has to follow as closely and continuously as possible all stages of the establishment and operation of the facility, the commissioning programme proposed by the applicant and all necessary changes must also be reviewed before permitting their implementation.

5.2.7. Approval Decisions

The review and assessment process will result in a number of regulatory decisions. At certain stages of the approval process the competent authority will take official actions resulting in the granting or the refusal of an approval.

5.2.8. Review of Approvals

Granted approvals can be re-examined by the competent authority in case of new information, which can affect safety considerations.

5.3. Regulatory Inspection and Enforcement

5.3.1. Objectives

The principal objectives are to ensure that:

- the persons responsible for the establishment and operation of the facility are competent;

- the required quality and performance of components, structures and systems are achieved;
- any deficiencies in equipment and procedures are corrected;
- the competent authority will be informed about experiences concerning operation, decommissioning;

5.3.2. Responsibilities

The competent authority is responsible for regulatory inspections according to a specified programme to assure that all safety related activities, systems as well as personnel competence satisfy the requirements according to approval.

Special regulatory inspections shall be carried out in case of abnormal occurrence requiring immediate investigation.

The competent authority should send a written warning or directive to the responsible organization in case of minor deviations from or violations of approval regulations specifying its nature and the necessary corrective actions including its timing.

In case of severe violations an immediate action will be taken by the authority requiring the operating organization to curtail activities.

In the event of very serious or chronic non-compliance with approval conditions or regulations the authority shall suspend or revoke the approval.

5.3.3. Inspection Functions

Regulatory inspections shall be performed in all areas of the regulatory responsibility with respect to siting, construction, operation, quality assurance programmes, reviewing results of periodic tests, preparing reports on regulatory inspections, checking emergency plans, etc.

5.3.4. Enforcement

The competent authority must have suitable power to enforce compliance with the relevant regulations and approvals.

5.4. Existing Facilities

Facilities built before the adoption of these recommendations should also be subject to regulatory control (and regular inspection) and the operating organization has to demonstrate to the competent authority that the facility achieves an acceptable safety standard.

6. Responsibilities of the Operating Organization

The operating organization responsible for the possession and use of the irradiator shall obtain from the competent authority approval, permit or authorization necessary for the acquisition, storage and use of the irradiator. The operating organization shall be responsible for the operation of the irradiator according to the given approval, permit or authorization.

The operating organization has the following responsibilities:

- To appoint radiation protection advisers in case of not having the necessary in-house radiation-technological expertise.
- To appoint at least two radiation protection officers with duties of ensuring the implementation of written administrative procedures and of assisting the organization to comply with the requirements of the approval.

- To assure that qualified operators ensure the safe operation of the facility having appropriate certificate of competence and approved training.
- To ensure regular staff training.
- To ensure personal monitoring (including also visitors), the availability of portable monitors and area monitoring with suitable instrumentation.
- To ensure regular testing and maintenance of safety instrumentation, interlock systems, control equipments. A programme shall be established and performed including weekly, monthly and semiannual tests.
- All tests, maintenance tasks, modification or changes to the irradiator shall be recorded.
- Regular maintenance for the whole facility shall be done according to the manufacturer's instructions. Modifications can be carried out only after approval from the competent authority and by qualified persons.
- To ensure safe operation clearly defined operational procedures laid down by the manufacturer and approved by the competent authority have to be followed.
- To ensure that all product positioning system components, product boxes or carriers continuously meet the design specifications.

7. Responsibilities of Other Relevant Parties

- The designers and manufacturers of the irradiation facilities shall ensure that the facilities meet the radiation safety objectives and any specific safety requirements of the competent authority.
- The importers and suppliers shall ensure that the facility is of safe design and that information on safe operating procedures is passed on to the operating organization.
- Constructors and suppliers shall ensure that their work does not compromise the safety aspects of the facility and they fully comply with the requirements of the designer and the manufacturer.

8. The Transport, Loading and Unloading of Sources

8.1. Transport

All transport of radioactive sources shall comply with the requirements of the IAEA Regulations for the Safe Transport of Radioactive Material and any existing national legislation [12].

A competent authority shall be appointed to set up and execute a programme for monitoring the design, manufacturing, testing, inspection, maintenance of packaging as well as to implement a system of documentation for the handling and storage of packages containing radioactive material by consignors and carriers.

In the case of transport accidents available recommendations should be used as guidance [13].

Due to recent events stricter regulations have been introduced with respect to the safe surface transport of radioactive sources. These regulations include stricter and more frequent reporting before, during and after the transport and the release of limited information about the shipment.

8.2. Loading and Unloading of Sources

Due to the potential hazard these operations shall be undertaken under close radiation protection supervision.

These operations may result in higher exposure to persons as compared to the conditions in normal operation of the facility, thus an evaluation is needed beforehand to keep the potential exposure as low as reasonably achievable.

Recent regulations allow the presence only of the skilled and designated staff at such operations.

9. Emergency Response Planning

The operating organization is responsible to prepare an adequate written emergency procedure (for each type of emergency foreseen) after having carried out a formal assessment of hazards [13]. In the case of an accident the operating organization shall initiate the emergency procedures and shall inform the competent authority and the radiation protection advisers.

Any incident shall be reported to the competent authority with a time schedule described in the approval and depending on the severity of the incident. Conclusions should be drawn to improve safety and to learn lessons with special attention to previous radiation accidents taken place earlier.

Extreme care should be exercised in the events of special problems with gamma facilities like the removal of damaged or leaking source, removal of contaminated material and actions under increased radiation levels.

10. Recent Safety Actions

Due to recent terrorist events a series of safeguards and threat advisories were issued to the major licensed facilities and new regulations were introduced worldwide. These security enhancements include e.g. measures to provide additional protection against vehicle bombs, water and land based assaults and tightened facility access control. The relevant regulatory commissions initiated programmes to reduce risks concerning the loss of radioactive materials and initiated the review of nuclear security programmes.

11. Conclusions

The radiation processing industry needs to review the existing protocols for facility safety and apply measures to assure that these regulations are fully implemented. The facilities must be designed with safety and all operational aspects of these facilities have to be taken into account. It is important to note that radiation safety at processing facilities has got so far a very good track record, since very few fatal accidents happened during the past four decades and almost all of them due to human errors. Consequently education and training is of basic significance to avoid accidents.

Taking into account the present sources of danger recent regulations concentrate on increasing the security by introducing closer personal control and by limiting the access of individuals. Stricter regulations have been introduced concerning the transport, as well as the loading and unloading of radioactive material.

References

[1] DU PLESSIS, T., OLIVIER, J.H.I., "Safety considerations and licencing of industrial gamma irradiation facilities", Procedures and modalities for the implementation of ISO codes in radiation sterilization, (AFRA Workshop, IAEA, Ghana, 1997), (unpublished paper).
[2] INTERNATIONAL ATOMIC ENERGY AGENCY, Radiation Safety of Gamma and Electron Irradiation Facilities, Safety Series No. 107, IAEA, Vienna (1992).
[3] INTERNATIONAL ATOMIC ENERGY AGENCY, Basic Safety Standards for Radiation Protection, Safety Series No. 9, IAEA, Vienna (1982).
[4] AMERICAN NATIONAL STANDARDS INSTITUTE, Safe Design and Use of Panoramic, Wet Storage Irradiators (Category IV), ANSI-N43.10-1984, New York (1984).
[5] INTERNATIONAL ATOMIC ENERGY AGENCY, Radiological Safety Aspects of the Operation of Electron Linear Accelerators, Technical Reports Series No. 188, IAEA, Vienna (1979).
[6] HEALTH AND SAFETY EXECUTIVE, An Introductory Guide to Programmable Electronic Systems in Safety Related Applications, HMSO, London (1987).
[7] INTERNATIONAL ORGANIZATION FOR STANDARDIZATION, Sealed Radioactive Sources – General Classification, ISO/TC 85/SC 2/WG 11N 31E, ISO, Geneva (1990).
[8] INTERNATIONAL ORGANIZATION FOR STANDARDIZATION, Sealed Radioactive Sources – Leakage Test Methods, ISO/TC 85/SC 2N 390, ISO, Geneva (1988).
[9] BRITISH STANDARDS INSTITUTION, Recommendation for Data on Shielding from Ionizing Radiation, Part 2: 1971, Shielding from X radiation, BS 4094, BSI, London (1988).
[10] NATIONAL COUNCIL ON RADIATION PROTECTION AND MEASUREMENTS, Radiation Protection Design Guidelines for 0.1 – 100 MeV Particle Accelerator Facilities, Rep. 51, NCRP, Washington DC (1977).
[11] INTERNATIONAL ATOMIC ENERGY AGENCY, Recommendations for the Safe Use and Regulation of Radiation Sources in Industry, Medicine, Research and Teaching, Safety Series No. 102, IAEA, Vienna (1990).
[12] INTERNATIONAL ATOMIC ENERGY AGENCY, Regulations for the Safe Transport of Radioactive Material, 1985 edition (As Amended 1990), Safety Series No. 6, IAEA, Vienna (1990).
[13] INTERNATIONAL ATOMIC ENERGY AGENCY, Emergency Response Planning and Preparedness for Transport Accidents Involving Radioactive Material, Safety Series No. 87, IAEA, Vienna (1988).

Radiation Inactivation of Bioterrorism Agents
L.G. Gazsó and C.C. Ponta (Eds.)
IOS Press, 2005

Dosimetry Systems for Radiation Processing

Arne MILLER
Risø High Dose Reference Laboratory, Radiation Research Department
Risø National Laboratory, DK-4000 Roskilde, Denmark
e-mail: arne.miller@risoe.dk, fax +45 4677 4959

Abstract. Several types of dosimeters are used in radiation processing, including calorimeters, thin radiochromic plastic films or sheets, free radical dosimeters, luminescent dosimeters and liquid dosimeters. These dosimeters, their properties and applications are briefly described in this paper.

Introduction

Radiation processing concerns many different applications, with the three main areas being sterilization of medical equipment, food preservation and polymer modification. National authorities first regulate the first two, and dosimetry is used to meet the requirements of these regulations. This is the case in radiation sterilization, where dosimetry is used in all stages of documentation of the process, both during Installation Qualification, Operational Qualification, Process Definition, Performance Qualification and Routine Monitoring [1, 2]. Other radiation processes do not require the use of dosimetry, and often product testing is used as the main test parameter. However, dosimetry is also in these cases useful for ensuring the documentation of a reproducible radiation process.

Several types of dosimeters are used for these purposes that utilize measurable chemical or physical changes in materials as a consequence of exposure to ionising radiation. The dosimeters include calorimeters, liquids in ampoules, plastics as films or thicker pieces, pellets and powders.

1. Calibration

The dosimeters used for controlling the radiation processes are considered routine dosimeters.

Because the response of *routine dosimeters* cannot, in general, be easily corrected for variations in environmental influence quantities, the dosimeters must be calibrated under the same conditions as those of eventual use. There are two main ways of achieving this,

a) Calibration in the irradiation plant and
b) Irradiation at a calibration facility followed by in-plant verification

In-plant calibration is achieved by irradiation of the dosimeters to be calibrated simultaneously with reference standard dosimeters supplied and measured by a calibration laboratory [3], preferably a laboratory with accreditation to prove its measurement traceability. It may be difficult to irradiate at exactly the doses preferred for the calibration, but it can often be overcome by irradiating the dosimeters to only part of the full irradiation cycle.

Calibration by irradiation at a calibration facility requires that steps be taken to verify that the response of the dosimeters irradiated in this way is not affected by the irradiation conditions of the plant where the dosimeter is going to be used. Without this verification step, systematic errors in the calibration may go undetected. Calibration verification can be carried out by comparing the doses measured by routine and reference standard dosimeters irradiated together in the irradiation plant at a few doses.

2. Standards

ASTM International has published a number of standards on dosimetry systems and dosimetry methods for radiation processing. In collaboration with the International Standards Organisation ISO most of these standards were published as ISO/ASTM standards [4]. The standards are listed below, and the highlighted standards are the ones used as references for the dosimeter systems described in this paper.

ISO/ASTM 51204 Practice for Dosimetry in Gamma Irradiation Facilities for Food Processing

ISO/ASTM 51205 Practice for Use of a Ceric-Cerous Sulfate Dosimetry System

ISO/ASTM 51261 Guide for Selection and Calibration of Dosimetry Systems for Radiation Processing

ISO/ASTM 51275 Practice for the Use of a Radiochromic Film Dosimetry System

ISO/ASTM 51276 Practice for the Use of a Polymethylmethacrylate Dosimetry System

ISO/ASTM 51310 Practice for the Use of a Radiochromic Optical Waveguide Dosimetry System

ISO/ASTM 51400 Practice for Characterization and Performance of a High-Dose Radiation Dosimetry Calibration Laboratory

ISO/ASTM 51401 Practice for Use of a Dichromate Dosimetry System

ISO/ASTM 51431 Practice for Dosimetry in Electron and Bremsstrahlung Irradiation Facilities for Food Processing

ISO/ASTM 51538 Practice for Use of the Ethanol-Chlorobenzene Dosimetry System

ISO/ASTM 51539 Guide for Use of Radiation-Sensitive Indicators

ISO/ASTM 51540 Practice for Use of a Radiochromic Liquid Dosimetry System

ISO/ASTM 51607 Practice for Use of the Alanine-EPR Dosimetry System

ISO/ASTM 51608 Practice for Dosimetry in an X-Ray (Bremsstrahlung) Facility for Radiation Processing

ISO/ASTM 51631 Practice for Use of Calorimetric Dosimetry Systems for Electron Beam Dose Measurements and Dosimeter Calibrations

ISO/ASTM 51649 Practice for Dosimetry in an Electron-Beam Facility for Radiation Processing at Energies between 300 keV and 25 MeV

ISO/ASTM 51650 Practice for Use of Cellulose Acetate Dosimetry Systems

ISO/ASTM 51702 Practice for Dosimetry in a Gamma Irradiation Facility for Radiation Processing

ISO/ASTM 51707 Guide for Estimating Uncertainties in Dosimetry for Radiation Processing

ISO/ASTM 51818 Practice for Dosimetry in an Electron Beam Facility for Radiation Processing at Energies between 80 and 300keV

ISO/ASTM 51900 Guide for Dosimetry in Radiation Research on Food and Agricultural Products

ISO/ASTM 51939 Practice for Blood Irradiation Dosimetry

ISO/ASTM 51940 Guide for Dosimetry for Irradiation of Insects for Sterile Release Programs
ISO/ASTM 51956 Practice for Thermoluminescence Dosimetry (TLD) Systems for Radiation Processing
ISO/ASTM 52116 Practice for Dosimetry for a Self-Contained Dry-Storage Gamma-Ray Irradiator
E 1026-95 Practice for Using the Fricke Reference Standard Dosimetry System
E 2232-02 Guide for Selection and Use of Mathematical Methods for Calculating Absorbed Dose in Radiation Processing Applications
E2303-03 Guide for Absorbed-Dose Mapping in Radiation Processing Facilities
E2304-03 Practice for Use of a LiF Photo-Fluorescent Film Dosimetry System

3. Dosimeters

A number of the most commonly used dosimeters are described in the following. They are:

- liquid systems in which the dosimeter solution is contained in a glass ampoule, and the response is measured by spectrophotometer or by electrical methods,
- solid systems which include:
- calorimeters that measure the temperature rise caused by absorption of energy,
- free radical systems that utilize a measurement of the concentration of the free radicals as the response signal,
- polymer films or sheets where the response is a measurement of the radiation-induced colour.

3.1. Ferrous Sulphate (Fricke) (ASTM E-1026)

This dosimeter is commonly used as a reference dosimeter, because its radiation chemistry is well understood. It consists of 0.001 M ferrous ammonium sulphate in an air-saturated, aqueous solution of 0.4 M sulphuric acid. Sodium chloride [0.001 M] is usually added to reduce the effects of impurities. The response is measured as the change in optical density at a wavelength of 303 nm.

The Fricke dosimeter has a typical lower dose limit of about 4 Gy and an upper limit of about 400 Gy. This modest range limits the use of the Fricke dosimeter for radiation processing, but its reproducibility and linearity of response make it a valuable dosimeter.

3.2. Cerric-cerrous(ISO/ASTM 51205)

This dosimeter consist of acidic aqueous solutions of tetravalent cerium ion salts, ceric sulphate hydrate [$Ce(SO_4)_2 \cdot H_2O$] or ceric ammonium sulphate used at different concentrations (0.2 to 50 mmol dm^{-3}) to cover different parts of the dose range 0.5 to 100 kGy. It is sufficient accurate and reproducible for reference dosimetry, however the purity of water and cleanliness of apparatus are important considerations since the reaction is affected by even trace quantities of organic impurities.

Another system, commonly referred to as the ceric-cerous dosimetry system, uses approximately equal initial concentrations of ceric and cerous ions to reduce the effects of organic impurities.

Analysis of the radiation-induced reduction of ceric ions can be carried out spectrophotometrically at the ceric absorption peak at 320 nm. The absorbance values for the dosimeter solution are too high for direct spectrophotometric measurement and require dilution with 0.8 N H_2SO_4 up to 100-fold to obtain absorbance values within the range of the spectrophotometer.

Ceric-cerous dosimeters can also be read potentiometrically, without the need to dilute the solution. This simplifies the readout procedure and allows the use of the ceric-cerous dosimeter for routine process monitoring. For potentiometric readout an electrochemical cell is required with a platinum electrode in an irradiated solution and another platinum electrode in an unirradiated solution separated by a porous junction.

The radiation-induced signal of the ceric-cerous dosimeter is stable and the dosimeter is used for reference dosimetry. However, is it also used as routine dosimeter, in particular using the potentiometric method of analysis.

3.3. Dichromate (ISO/ASTM 51401)

The most commonly used formulation of the dichromate dosimeter is a mixture of potassium and silver dichromate in 0.1 M perchloric acid. The absorbed-dose range depends on dichromate concentration; for example, a solution containing 0.5 mM $Ag_2Cr_2O_7$ can be used between 2 and 10 kGy. The operating range can be modified to 5 to 55 kGy by the addition of 2.0 mM $K_2Cr_2O_7$.

The dichromate dosimeter is measured by spectrophotometry in terms of the decrease in concentration of the dichromate ion. The solution containing 0.5 mM $Ag_2Cr_2O_7$ can be measured at the peak of the absorption spectrum at 350 nm, while the solution containing, in addition, 2.0 mM $K_2Cr_2O_7$ can be measured at an inflection of the absorption spectrum at 440 nm.

Also this dosimeter is very stable and it is used for reference dosimetry, but usually not for routine dosimetry. Generally, methods using spectrophotometry that involves tranfer of the dosimeter solution to the measurement cuvette is usually consider too cumbersome to be practical.

3.4. Ethanol-Chlorobenzene (ISO/ASTM 51538)

Ethanol chlorobenzene dosimetry is based on the radiolytic formation of hydrochloric acid in irradiated aqueous, ethanolic solutions of chlorobenzene, together with a method of hydrochloric acid measurement. A number of analytical methods are available for the measurement of HCl in ethanol. Depending on the readout method, the system can be used as a reference dosimeter or as a routine dosimetry system.

The most often used method of measurement for routine dosimetry is a high-frequency conductivity measurement, often referred to as oscillomerty. It is a non-destructive method; the dosimeter ampoule is not broken, but inserted into a holder in an electric measurement circuit. Using this method the dose range is from a few kGy to more than 100 kGy, depending on the chemical formulation of the dosimeter solution.

3.5. Calorimeters (ISO/ASTM 51631)

Calorimetry is the most fundamental method for measuring absorbed dose. The response of a calorimeter such as used as primary standards in national laboratories related to the definition of absorbed dose D, as the quotient of the mean energy, de, imparted by ionizing radiation to matter of mass dm, and measured as a rise in temperature ΔT.

The calorimeters that are used in radiation processing for measurement at dose at electron accelerators are, however, relatively simple. They may consist of an absorber placed in heat-insulating plastic foam and with a temperature-sensitive sensor (thermistor,

Table 1. Commercially available.

Dosimeter Type	Nominal thickness (mm)	Analysis wavelength (nm)	Absorbed dose range (kGy)
Red 4034	3	640	5 to 50
Amber 3042	3	603, 651	1 to 30
Radix RN15	1.5	315	5 to 50
Gammachrome YR	1.5	530	0.1 to 3

thermocouple or platinum resistor) placed in the absorber for temperature measurement. They can be influenced by external factors (for example, heat from warm zones at electron accelerators), and they therefore have to calibrated against reference dosimeters.

Calorimeters for radiation processing have been constructed for a wide range of specific applications and the calorimeter absorbers have been made from several different materials, including graphite, water, metals and polymeric materials e.g. polystyrene.

'The calorimeters are usually measured "off-line", i.e. the temperature is measured before and after irradiation only, and this limits the sensitivity compared to what is possible for on-line measurements. Typical dose ranges for different absorbing materials are:

Graphite: 1.5 – 15 kGy
Polystyrene: 3 – 40 kGy
Water: 5 – 50 kGy

These dose ranges are based on a minimum temperature increase of 2°C and a maximum calibrated temperature of 50°C .

3.6. Alanine (ISO/ASTM 51607)

Electron paramagnetic resonance, EPR (or electron spin resonance, ESR) spectroscopy of the amino acid alanine is used for dosimetry. Solid dosimeters in the form of pellets, rods, and films are produced from polycrystalline α-alanine powder and an additive binder. Pure alanine powder without further processing has also been used.

Generally, the dose range of the alanine-EPR dosimetry is from 1 Gy to 100 kGy. The actual dose range of a specific system depends on the alanine dosimeter type and the sensitivity of the measurement instrument.

The EPR signal is very stable over a time scale of months and alanine dosimeter is therefore the dosimeter most often chosen by reference laboratories as transfer dosimeters. The dosimeters repeatedly.

3.7. PMMA (ISO/ASTM 51276)

Polymethylmethacrylate (PMMA) both dyed and colourless, have radiation-coloration properties that make the material useful as dosimeters. Sheet material with thickness in the range of 1 to 4 mm is cut into plaques of a convenient size for spectrophotometric measurement, for example 12 x 38 mm. The dosimeters are sealed in pouches in order to protect them against changes in humidity and to protect the surfaces for optical measurement.

Commercially available dosimeters collectively spans a dose range of 100 Gy to 100 kGy as given in Table 1.

PMMA dosimeters are used as routine dosimeters mainly in gamma irradiation facilities, however the response of PMMA dosimeters show some degree of instability after irradiation and they are therefore not suitable for reference transfer dosimetry.

Table 2. The nominal dose ranges for PCR dosimeters.

Dosimeter Type	Nominal thickness (micron)	Analysis wavelength (nm)	Absorbed dose range (kGy)
FWT60	50	605	1.5 to 50
FWT60	50	510	15 to 100
WinDose	20	554	3 to 100
GAFchromic	1.5	530	0.01 to 100

3.8. Radiochromic Films (ISO/ASTM 51275)

The radiochromic dye film dosimeters (RCD) are of freestanding or coated polymeric thin films, which are colorless prior to irradiation. They are usually supplied as small squares, strips, or long rolls or sheets that can be cut into a convenient size for dosimetry. For routine (and reference) dosimetry, these dosimeters are usualy proteced in moisture-proof pouches, while for dose mapping the dosimter are used without protection.
The nominal dose ranges for commercially available RCD dosimeters listed in Table 2 shows that collectively theses types of dosimeters cover a relatively broad range, namely 0.01 to 100 kGy.

The response of the RCD dosimeters is relatively stable after irradiation and they are used for reference dosimetry, although the most common use is routine dosimetry, both at gamma and electron irradiation facilities.

3.9. CTA (ISO/ASTM 51650)

Cellulose triacetate (CTA) film containing a plasticizer, triphenylphosphate (TPP), is used for dosimetry in the dose range frm approximately 10 to 300 kGy. The spectrophotometric response is measured in the UV region, typically at 280 nm, which is at a steep slope of the absorption spectrum. The response function is almost linear over the useful dose range.

The 125 micron thin dosimeter film is uslly provided as rolls at lengths of 100 m. It is mainly used for dose mapping at electron beam facilities.

4. New Developments

The literature contains a wealth of descriptions of potential dosimeter systems or of materials that have been tested with respect to their dosimetric properties. This paper does not intend to give a complete evaluation of all these, but a few new developments are described. The omission of many others is not indicative or their potential usefulness – or the lack of it.

Luminescence phenomena in several materials are used for dosimetry, and recently photo-luminescence of LiF has been applied in high-dose dosimetry (ASTM E2304). The dosimeters consist of LiF powder suspended in a polymeric film matrix and cut into sheets of typically 1 x 2 cm^2 sizes. These dosimeters that are marketed as "Sunna" dosimeters can measure dose in a range from approximately 50 Gy to more than 100 kGy.

Another development concerns not the dosimeters, but their measurement technique. Optical measurements of many dosimeters are usually done at spectrophotometers, but recently it was shown that a simple office scanner coupled with appropriate software can be used for measurement of thin radiochromic dosimeters with good reproducibility [5]. The software called "RisøScan" is particular suited for dose mapping measurements and can quickly identify and measure minimum and maximum doses.

5. Conclusion

Dosimeter systems provide the needed documentation for correct execution of radiation processing applications, in particular radiation sterilization. These dosimeter systems are well documented in the literature and in international standards, and they can fulfil the required tasks. New systems are being developed that can further improve the capabilities in terms of reduced measurement uncertainty and increased spatial resolution of the dose measurement.

References

[1] ISO (1995). ISO 11137, Sterilization of health care products – Requirements for validation and routine control – Radiation sterilization. International Organization for Sterilization, C.P. 56, CH-1211 Genève 20, Switzerland.

[2] CEN (1994). EN 552, Sterilization of medical devices – Validation and routine control of sterilization by irradiation. European Committee for Standardization, Rue de Stassart 36, B-1050 Brussels, Belgium.

[3] CIRM 29, (1999) Miller and Sharpe, Guidelines for the Calibration of Dosimeters for use in Radiation Processing. National Physical Laboratory Teddington, TW11 0LW, United Kingdom.

[4] ASTM and ISO/ASTM standards are available from: ASTM International, 100 Barr Harbor Drive, PO Box C700, West Conshohocken, PA 19428-2959, USA.

[5] J. Helt-Hansen and A. Miller (2004), RisøScan – A new software for dose measurement. Rad Phys Chem, to be published).

Process Control of Radiation Treatment

Corneliu C. PONTA
IRASM Department, Horia Hulubei National Institute for Physics and Nuclear Engineering, 407 Atomistilor, Bucharest - Magurele, PO Box MG-6, Romania

Abstract. A wide spread industrial application of radiation treatment is sterilization. The control of any process is a responsible and documented action for safe release of goods to the end-user. In radiation sterilization the control strategy has a validation step followed by routinely monitoring of validated parameters. Validation procedure establishes the sterilization conditions of a specific product treated in a specific irradiation facility. During validation the bioburden level before sterilization and radiation sensitivity of microorganisms found on the items (D_{10} value) are measured. They contribute together with an agreed SAL (Sterility Assurance Level) to choose the irradiation time direct related to the sterilization dose. Parametric release to the market is accepted if routine process proceeded in validated conditions. For radiation inactivation of biological weapon, validation concept can not be applied in an orthodox way mainly because the bioburden level can not be measured. ALARA (As Low As Reasonable Achievable) concept is a possible approach to establish a strongly motivated treatment dose.

Introduction

Manufacture - Control - Release to the market. These are the main important steps from the producer to the end-user for any product. In this chain, the word "control" usually means "verification". After manufacturing (or after one single manufacturing step), the product (in the final form or in an intermediate one) is tested versus a specification and if fits to it, is released accompanied by a conformity certificate (or equivalent).
In some technological processes this control strategy can not be followed, because by verification action the tested state of the product is altered. Such processes, whose results can not be inspected but in a destructive way, are denominated as "special" in EN standards' terminology.

"*Sterilization* is an example of a special process because process efficacy can not be verified by inspection and testing of the product" [1]. Additionally, SAL concept, defining "sterile" term on a probabilistic base, makes testing by sampling irrelevant.

For a special process like radiation sterilization the control strategy starts with a first step called process validation. The microbiological parameters of untreated products are measured. The irradiation dose and technical specification able to assure a successful treatment are evaluated. For routine treatment, a few numbers of validated parameters are monitored and documented. The "parametric release" to the market is closely related to this control strategy. Parametric release means that the treatment certificate is accepted if the process proceeded in validated conditions.

In this control strategy, called later "validation strategy", the word "control" has an interactive meaning because it includes a feedback relationship. Process control means irradiation process control also irradiation efficiency control. It also shares responsibilities between the manufacturer and radiation facility. The validation strategy may be applied if the products are manufactured under GMP rules.

This paper presents the control procedure of radiation sterilization. It also tries to evaluate how it can be adapted to the irradiation inactivation of bioterrorism agents.

1. Validation of Radiation Sterilization

1.1. Fundamentals

Radiation treatment is a syntagma used for the treatment with ionizing radiation.

An ionizing radiation has enough energy to break a chemical link and change a molecule. This property is used in radiation treatment.

- The effect of radiation treatment depends on the amount of transferred energy. The energy transfer is evaluate by the "absorbed dose" - D, later also called "dose" or "radiation dose"; the unit of D is called "gray" (Gy); 1Gy = 1J/kg; D measuring systems are called "dosimetry systems".
- In microbiological evaluation a living microorganism is considered the one capable to divide and being able to form a colony. The living cell is called "colony forming unit" (CFU).
- Basic effect in radiation of a single microorganism is DNA modification. Consequently the cell can not form a colony anymore and the microorganism is considered dead.
- "Inactivation of a pure culture (a population - our note) of micro-organisms by the physical and/or chemical agents ...approximates to an exponential relationship; inevitably this means there is always a finite probability that a micro-organism may survive regardless of the extent of treatment applied" [1]. This citation from EN 552 mentions a basic natural fact. Consequences are the following:
- In any exponential variation a constant may be identified that is typical for the specific relationship. By radiation of a certain microorganism population, the graphic of survival "CFU versus D" (Fig.1) emphasizes D_{10} value as a constant typical for radiation of that specific microorganism. D_{10} is the irradiation dose that inactivates 1/10 of the total living population.
- One can not obtain by any means an absolute sterility status of a batch. A certain risk has to be assumed in establishing a sterilization procedure. This risk is quantified by the "Sterility Assurance Level" - SAL. It has a mean and is applied only for great numbers of similar items treated in the same conditions - a batch.

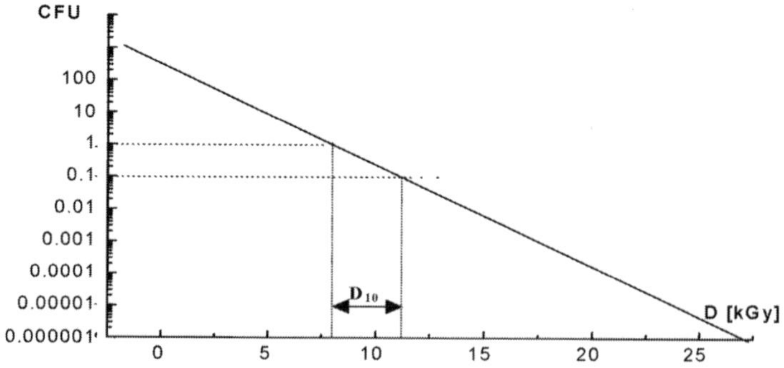

Figure 1. A theoretical graphic of survival.

1.2. Establishing the Treatment Dose. Microbiological Validation of the Sterilization Dose

To have the legal right to label "sterile" a medical device, one has to document a sterilization procedure capable to assure a SAL = 10^{-6}. It means that one from a million items could be non-sterile. Both [2] and European Pharmacopoeia mention this SAL value as an accepted risk. Health authorities may approve exceptions.

Microbiological validation of the sterilization dose essentially means to establish and to document the necessary dose to attain a convened SAL. A specific validation file is required for any product or product category. A revalidation is asked from time to time.

Historically 25 kGy was considered the sterilization dose for medical devices. There was no obligation for a microbiological validation of any particular case.

However 25 kGy was chosen taking into account SAL = 10^{-6}. Two premises stated behind: a) the bioburden level on an item is no higher than 10^2 CFU; b) the bioburden consists only in Bacillus pumilus. In that time Bacillus pumilus was known as the most radiation resistant microorganism, having D_{10} = 3.1 kGy. The bioburden level of 10^2 is justified by being the maximum accepted level for a non-sterile medical device manufactured in conformity with GMP rules.

In the above conditions the sterilization dose resulted considering a dose 8 times more than D_{10} value (to attain 10^{-6} starting from 10^2):

$$8 \times D_{10} = 8 \times 3.1 \text{ kGy} \sim 25 \text{ kGy}$$

25 kGy was a sterilization dose evaluated by overestimation, supposed to be safe in the worst imaginable case.

Meantime the initial premises changed. Microorganisms more radiation resistant than Bacillus pumilus were discovered. The health authorities claimed for higher sterilization doses. Technical progress improved the manufacture conditions. The theoretical value of 10^2 CFU per non-sterile item can not be sustained anymore. The manufacturers claimed for lower sterilization doses.

The compromise accepted by both parts was the establishing of sterilization dose by a validation procedure particular to each product type manufactured in the same conditions.

For dose setting the applied rationale is the same as in 25 kGy establishing.

Three values have to be known: bioburden level, D_{10} value and the agreed SAL. Two of them - bioburden level and D_{10} value - *have to be measured* by a competent microbiological laboratory. The third - SAL - *is mandatory by convention.*

Details on microbiological evaluation are out of the scope of this paper.

1.3. Delivering a Dose in Industrial Facilities

Ionizing radiation currently applied in industry is gamma radiation provided by radioactive isotopes Co-60 and Cs-137, or electron beams produced by accelerators.

Gamma radiation is of electromagnetic nature. It is delivered in all directions and can not be focused. The energy transfer from the radiation to the exposed matter is weak. As a result gamma ray is highly penetrant making possible the benefit of irradiation large volumes in the same time. On the other hand it is not possible to use all energy provided by a gamma source.

Accelerated electrons are delivered in a focused beam having a certain direction. Being charged particle, electrons interact strongly with the exposed matter. A penetration depth can be defined for electrons of certain energy. In between this depth the electrons lose all energy and are annihilated.

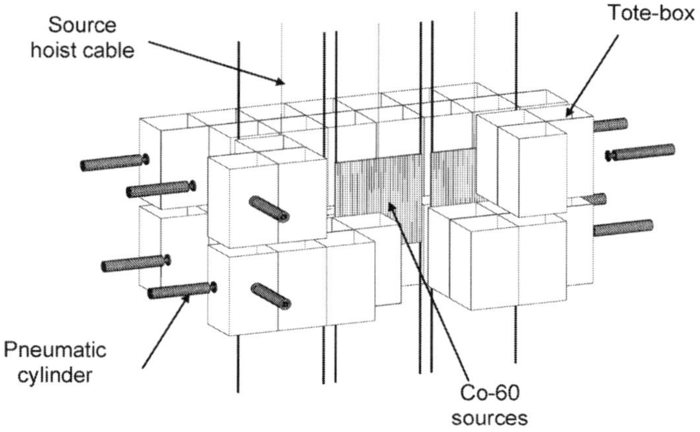

Figure 2. Irradiation geometry in IRASM tote-box type facility.

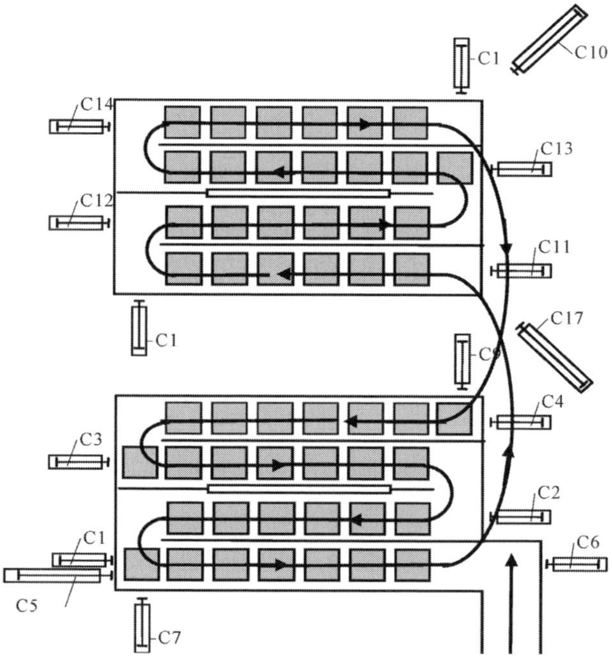

Figure 3. Conveyer pass.

Radiation facilities are designed taking into account the interaction particularities of specific radiation type. Both facility types have to solve the non-homogeneity of radiation field, in other words dose rate non-homogeneity.

Radioisotopes are placed in gamma irradiators to form either a plaque or a cylinder. The radiation field is provided in a protected area called irradiation room. Here the goods loaded in tote-boxes (carriers or pallets) are moving in a definite number of steps following a determined geometry. Fig. 2 and 3 present the irradiation geometry at IRASM facility.

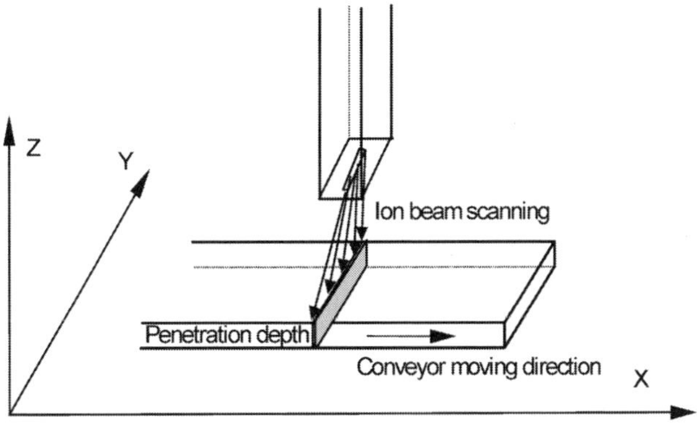

Figure 4. Radiation geometry in accelerators.

The time spent in each step is known as "cycle time". In each cycle time the goods accumulate a dose increment. The increment depends on distance between the source and the tote (carrier, pallet) also on the package density. The non-homogeneity depends on the conveyor design related to source geometry for the same package density. The accumulated dose is in direct relation with the overall irradiation time.

The non-homogeneity of radiation field delivered by accelerators is due to the specific interaction of charged particles. In accelerators the goods are treated in thin packages having actually two significant dimensions. The third one is much smaller and depends on the penetration depth. The electron beam is rapidly moving to scan a line. A high-speed conveyor transports the packages in a direction perpendicular to the scanning line (Fig. 4). By this radiation arrangement non-homogeneity of radiation field can be limited. The dose rate depends on the electron energy and the beam current. Scan width and scan uniformity is also involved. These parameters defining the radiation field can be adjusted. They have to be correlated with the density of the treated package. When all of them are fixed for an acceptable non-homogeneity, radiation dose directly depends on conveyor speed only.

The biological effect of radiation depend on accumulated absorbed dose, regardless the radiation type, dose rate or interruption in delivering the dose. Continuous irradiation procedures are possible. Facilities' flexibility in adapting radiation conditions to a given case makes radiation a safe and reliable tool for industrial sterilization.

1.4. Dose-mapping

Regardless of facility type - gamma or accelerator - non-homogeneity in delivering the dose can not be avoided. Items in a collective box (gamma facilities) or parts of a single package (accelerators) will receive different doses in the same time.

"The dose mapping should provide information on the distribution and homogeneity of the dose over the irradiation field and at various known depth of an absorber of known density. This is performed by determining absorbed dose levels at points within a stack of homogeneous absorber sheets where dosimeters are placed on the surfaces of the sheets and between the sheets" [1].

Fig. 5 presents a dose mapping inside a tote-box at IRASM facility - Romania.

The non-homogeneity is quantified by the overdose ratio. It is defined as D_{max} / D_{min}.

D_{min} is necessary to calculate the irradiation time. Obviously D_{min} has to be greater or at least equal to Sterilization Dose.

D_{max} is important for evaluation of radiation resistance of treated and packaging materials.

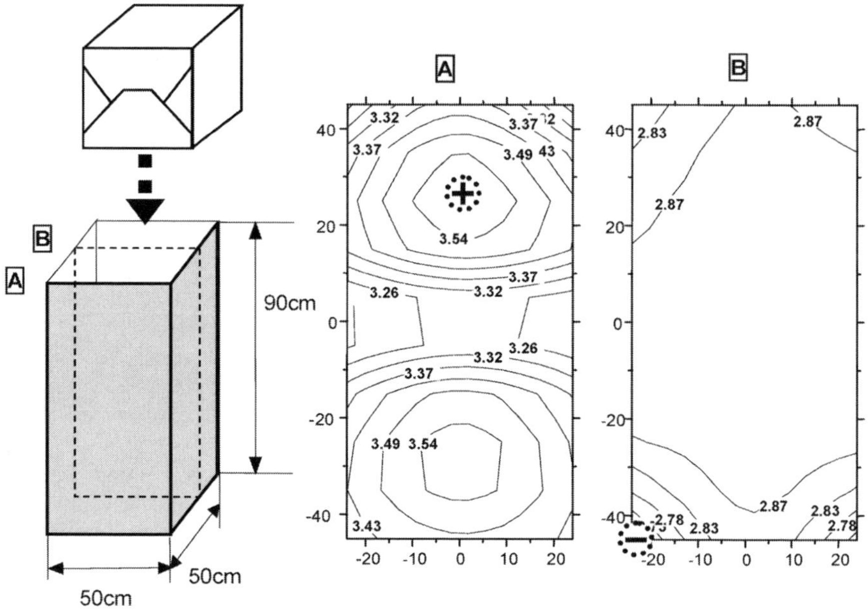

Figure 5. Dose mapping D_{min} ⊖ and D_{max} ⊕ are positioned in different planes of the tote-box.

It is important to notice that for a given stuff in an established packaging arrangement, exposed in a given irradiation field, D_{min} and D_{max} are not independent variables.

To find out the dose non-homogeneity a three-dimension dose mapping is required inside a tote or a package. The values and the positions of D_{min} and D_{max} are established.

Dose mapping is an important part of validation procedure. It is also useful for the routine control.

1.5. Validation Brief Summary

Sterilization validation is mandatory for each product or product category. The validation establishes the conditions to perform a properly and consistently radiation sterilization.

Important values obtained in the validation procedure are:

- Bioburden level before sterilization;
- Sterilization dose;
- Operation parameters to obtain sterilization dose: operation time for gamma irradiators or accelerator parameters: electron energy, beam current, scan width, scan uniformity and conveyor speed;
- Loading pattern for which operation parameters act efficiently;
- D_{min} and D_{max} positions in the irradiation container.

Manufacturer and radiation facility, also the involved microbiological laboratory shares responsibilities during the validation.

1.6. Routine Control of Radiation Sterilization

Parametric release to the market is accepted if routine process proceeded in validated conditions.

The radiation facility is responsible for keeping the routine operation parameters in the limits established in the validation stage. The routine control of radiation dose is made by dosimeters placed in D_{min} position or in a position mathematically related to D_{min}.

It is manufacturer responsibility to keep the bioburden level and the loading pattern in the limits took into account at validation step.

2. Sampling as Control Method in Radiation Disinfection

Disinfection also called microbial control is another application of radiation treatment. Its target is to obtain a pathogen-free product and/or to minimize bioburden level in foodstuff, cosmetics or pharmaceutical row materials.

For the use of this paper the control strategy is called "sampling strategy".

The radiation control methods are the same as in the sterilization case.

Control of treatment efficiency is performed by sampling followed by microbiological tests of a chosen number of treated stuff. The treatment dose as main process parameter is not established by microbiological validation. It would be useless because these products have large variation of bioburden level. Other considerations are used for this purpose. In the typical case of food treatment health authority issues a license mentioning the upper limit of the treatment dose. Parametric release is not accepted. The tested goods are sacrificed. Quarantine is necessary for the test completion.

This rather simple control philosophy derived from the fact the radiation disinfection is thought as a preventive procedure. The treatment dose is much smaller than currently used sterilization dose. In Europe it is up to 10 kGy.

3. A Challenge - Control Strategy in Radiation Inactivation of Bioweapons

Previous considerations emphasized the control means both radiation process control and radiation efficiency control. Inactivation of bioweapon by radiation has to be approached in the same way. Radiation process control is the same regardless the treatment intention. *Efficiency control needs a special strategy adapted to the case.*

The main problem is to establish the proper dose. The two mentioned applications have different rationale for this purpose.

- In sterilization the proper treatment dose is validated during a complex process. One may say the "start point" (bioburden level), the "end point" (SAL) and the "stroke" (D_{10}) are needed to apply a scientifically derived rationale. Bioburden level and D_{10} value are measured during the validation stage. SAL value is considered by convention (10^{-6} for many product categories). The treated goods are parametric released.
- In foodstuff disinfection the treatment dose is not derived from a calculation. It is imposed as a maximum value by health considerations. The value is based by less accurate reasons. Reasonable maximum pathogens' concentration on the product (as the start point) is among the presumptions. After the treatment, the goods are released only after testing of a definite number of samples.

The radiation procedure proper for inactivation of bioterrorism agents eventually placed in packages has to permit parametric release based on an indisputable efficiency.

In sampling strategy parametric release is not accepted.

Let's examine in what way the rationale used in validation strategy fits to the case.

Parametric release is based on clearly defined premises: repeatability of bioburden found on the item and the identification of its level performed before treatment. It is obviously not the case. No one can measure the bioweapon level contaminating a not identified stuff. *The good question is how to substitute the measurement of bioburden level.*

To have a high efficiency procedure SAL concept was developed for sterilization case. It could be applied if its value is convened. *To convene a safe SAL is the challenge.*

D_{10} value for a specific population is known or can be measured.

Trying to solve the raised questions we may use ALARA concept.

ALARA is the acronym of expression As Low As Reasonable Achievable. It is a philosophical concept involving the *scientifically* common sense for establishing safety measures or intervention levels against terrifying phenomena or threats hard to predict: nuclear accidents, earthquakes. Designing the safety measures / systems in conformity with ALARA, reasonable limits and resources are accepted. This concept can be identified in establishing 25 kGy as sterilization dose also in sampling strategy.

The following is just an exercise for establishing the treatment dose based on ALARA concept. Considerations and taken values where this concept is applied are indicated in brackets.

Data from biotechnological production of microorganisms useful in agriculture are considered. In a bioreactor the vegetative form is multiplied and converted into spores. They are concentrated by centrifugation, dried and mixed with a carrier.

- *Evaluating the reasonable achievable bioburden level that may intentionally contaminate mail*

Basic assumptions:

- the volume extracted from bioreactor contains only living spores (ALARA);
- the spore volume = $1\mu m^3 = 10^{-12} cm^3$; $1 cm^3$ will contain 10^{12} living spores;
- the spore survival yield at drying = 10^{-6}; chosen value is 10^{-5} (ALARA);
- the spore concentration in carrying substance is $1\% = 10^{-2}$ (as is usually the case for manipulation reasons); chosen value is $10\% = 10^{-1}$ (ALARA);
- volume of contaminating substance in one envelope = $1 cm^3$ (ALARA);

Than we can derive the reasonable achievable bioburden level: living spore quantity (CFU) in one envelope = $10^{12} \times 10^{-5} \times 10^{-1}$ = **10^6CFU**

- *D_{10} value*

We chose Bacillus anthracis to complete this exercise. Scientific information about its D_{10} value is scarce. **D_{10} value** = **3.35 kGy** is used here [3].

- *Establishing the SAL value and the treatment dose.*

There are many possible comments on this subject.

It seems that one single spore or a small spore number of Bacillus anthracis is tolerated by the human organism. Medical doctors established $LD_{50} = \sim 10^4$ CFU for lung infection [4]. Taken this observation into account we do not need to choose a SAL value.

It seems reasonable to choose eventually 1 CFU as "end point" of treatment (ALARA).

In this case the treatment dose = 3.35 x 6 = 20.1 kGy

If the above judgement is not acceptable, a reasonable number of envelopes contaminated (and treated) in the same time have to be considered. We can choose 10^3 (ALARA). The derived SAL is 10^{-3}. It means to accept the risk that 1 envelope from contaminated 1000 will eventually remain with 1 CFU after treatment.

In this case the treatment dose = 3.35 x 9 = 30.15 kGy

Note: A SAL = 10^{-6} is in overestimation difficult to accept in Anthrax case. It would mean to accept the possibility of having one million contaminated packages in the same time.

Similar rationale could be applied to each specific microorganism which is a potential bioterrorism agent. Having a list with the inactivation dose for each of them, ALARA concept could be applied again to decide a treatment dose safe in any case.

The author does not pretend to have the scientific and technical knowledge to define the reasonable achievable levels chosen in the above exercise. He only claims for the benefit of using the above rationale in establishing the efficient treatment dose.

4. Conclusions

Bioweapon is a terrifying threat comparable with nuclear accidents or earthquakes.

There are important reasons to consider radiation a potential tool for inactivation of those bioterrorism agents based on microorganisms. This tool can manage the risk. Radiation treatment proved to be a reliable and efficient procedure for industrial sterilization and microbial control. Routine process control is simple and well established. Facilities in large numbers are spread over the world. No important adaptations seem to be necessary.

For an efficient treatment most important parameter is the treatment dose. Neither validation nor sampling strategy can be used in an orthodox way for establishing the efficient radiation dose for bioweapon inactivation.

ALARA concept is proper to solve the problem because the risk could be evaluated using scientifically and technical knowledge. A strong motivation in evaluation process will avoid temptation to overestimate the risk. Otherwise the treatment costs could become unbearable with consequences hard to predict.

References

[1] EN 552: 1994, *Sterilization of medical devices - Validation and routine control of sterilization by irradiation.*

[2] EN 556: 1994, *Sterilization of medical devices - Requirements for medical devices to be labelled "Sterile".*

[3] Niebuhr S.E. and Dickson J.S., *Destruction of Bacillus anthracis strain Sterne 34F2 spores in postal envelopes by exposure to electron beam irradiation,* Letters in Applied Microbiology, vol. 37, issue 1, pg. 17, July, 2003.

[4] *** *Anthrax,* MCT Internal Report A 151, November, 2001 (in Romanian).

Radiation Inactivation of Bioterrorism Agents
L.G. Gazsó and C.C. Ponta (Eds.)
IOS Press, 2005

Dose Setting Procedures for Radiation Sterilization

Iwona KALUSKA and Zbigniew ZIMEK
Institute of Nuclear Chemistry and Technology
Department of Radiation Chemistry and Technology
Dorodna 16, 03-195 Warsaw, Poland

Abstracts. The manufacturer of the product has responsibility for the quality of the product including the selection of the appropriate sterilizing dose. Several approaches to select the dose can be used depending on the batch size and device bioburden level.

Introduction

Radiation sterilization of medical devices has been used since 1953 when Ethicon began to sterilize surgical sutures. Sterilization is the final manufacturing operation that can significantly affect the safety and effectiveness of a finished device. After many years of investigations the dose of 25 kGy became established and accepted by many regulatory authorities in Europe as a minimum irradiation dose. However, there were also another suggestions. Doses of 35-45 kGy were used in Scandinavian countries according to evidence of higher radiation resistance of some environmental isolates. On contrary, the US authorities have allowed to use lower doses based on product specific dose setting studies.

1. Sterilization Dose Selection

Standards concerning radiation sterilization appeared in mid 90's – EN 552 [1] and ISO 11137 [2]. As far as sterilization dose is concern both of these standards allow one of two possible approaches to be used for it selection:

a) selection of sterilization dose using either
   1. bioburden information or
   2. information obtained by incremental dosing
b) selection of a sterilization dose of 25 kGy following substantiation of the appropriateness of this dose.

However, in EN 552 is added one more sentence at the ends of these two approaches– the sterilizing dose chosen shall be capable of achieving compliance with EN 556 [3].

According to EN 556, which describes the requirements for medical devices to be labeled sterile, the product is sterile only when sterility assurance level [SAL] achieves 10^{-6}.

ISO supports the notion of dual SAL's for sterile health care product depending on product use: SAL of 10^{-3} for topically applied medical devices and SAL of 10^{-6} for implantable devices.

In Annex B of ISO 11137 two setting methods are described for selecting product-specific sterilizing doses, Method 1 and Method 2.

1.1. Method 1

Method 1 depends upon experimental verification that the response of the product microflora to radiation is equal to or greater than that of a microbial population having a standard distribution of resistance (SDR). The SDR is a hypothetical distribution based on the measurements of the radiation resistances of selected microbial isolates, comprising a series of increasing D_{10} values and associated probabilities of occurrence.

The verification dose is the dose that will reduce a microbial population having a standard distribution of resistance to a level that gives a one in 100 chance of occurrence of non-sterile product.

Method 1 consists of 5 stages of action. In the first stage the intended sterility assurance level and samples of products units are selected. There are needed at least 10 items from each of three independent production batches.

In the second stage individual bioburden of each of 30 items is determined. If products are too big for sterility test, a sample item portion (SIP) is to be used instead. When bioburden is determined for each of the three batches the overall batch average can be calculated. In this stage it is important to compare the range differential between the highest and lowest recovered number in each group of ten items. If there is an individual bioburden result, which is two or more times greater than the group average, this value should be taken to further action. If there is no big difference in bioburden results than overall batch average is calculated. In the third stage the verification dose is selected from the dose table B1given in the Annex B -ISO 11137 based on the value of calculated bioburden in the second stage. In the next stage the 100 final products are irradiated with the dose selected in the third stage. The crucial thing is to deliver dose, which can vary only by +10%. If there is any doubts that delivered dose does not meet the specification, it does not make sense to proceed to the next stage. The verification dose experiment has to be repeated with the next 100 samples. But if dose was delivered precisely than procedure can be continued, it means the sterility test of 100 units should be performed. If there are no more than two positive sterility tests from the 100 tests carried out, the verification procedure is passed. In the last stage the sterilization dose can be established. Again using table from Annex B ISO 11137 for the calculated bioburden, necessary dose to achieve selected SAL can be read off. To summarize Method 1, it should be pointed out that this method could be used for any batch sizes and any production rate. However, the bioburden is limited to 1 million and there are needed at least 130 sample units for testing.

1.2. Method 2

Method 2 uses the results of tests of sterility performed on samples of products that have been irradiated with the series of incremental doses to estimate the dose, at which a SAL of 10^{-2} is reached. The results of the sterility tests are used to make an estimate of the D_{10} value and this estimate is use for extrapolation to SAL's below 10^{-2}. There are two procedures for validation of method 2. Method 2A can be used for products from any normal manufacturing process and Method 2B is devoted only for products with a consistent and very low bioburden. It is why in Method 2B only whole products can be used while in Method 2A either an entire product or a portion thereof. In Method 2 at least 640 product samples are required because at the beginning of the procedure 20 product units from each of three batches are irradiated in a series of not less than nine doses, increasing in 2 kGy increments. The steps in Method 2B, which refers to the products with consistent and very low bioburden, are similar to Method 2A, but the incremental dosing units increase in 1 kGy increments instead of 2 kGy and after irradiation at any of the incremental doses, the number of positive tests should not exceed 14.

Method 1 and Method 2 can be also used for substantiation of sterilization dose of 25 kGy, provided that the sterilizing dose established using these methods is less than 25 kGy. To perform substantiation of 25 kGy with mentioned above methods is quite expensive.

2. Method VD_{max} – Substantiation of 25 kGy Dose

J. Kowalski and A. Tallentire proposed alternative approach for demonstration the effectiveness of 25 kGy [4]. This method, called VD_{max} , based on the SDR of Method 1, can be used for batches of any size and manufacture at any production rate, which average bioburden is less than 1000 colony forming units (cfu) per device. This approach is valid for sample size from 10 to 100 units, corresponding to SALs of 10^{-1} to 10^{-2}. The approach aims to:

- Link directly the outcome of the verification dose experiment with the attainment of an SAL of 10^{-6} at 25 kGy;
- Provide the maximal verification dose that is consistent with this linkage; and
- Provide a level of conservativeness at least equivalent to that built into the standards distribution of resistances (SDR).

The procedural elements of VD_{max} methods are as follows:

- Performing a bioburden demonstration;
- Based on the knowledge of the number of product units to be used for verification dose experiment, a simple calculation is conduct to obtain the maximal verification dose characteristic of the resultant bioburden estimate;
- Performing a verification dose experiment;
- Based upon the outcome of the verification dose experiment the substantiation of a 25 kGy is accepted or rejected.

VD_{max} preserves the conservative aspects of the resistance characteristics of the SDR, but is more accurate for low bioburden products. VD_{max} method will be included in the revision of ISO 11137.

3. Dose Selection for Tissue Allografts Sterilization

Mentioned above standards concern medical devices and the question is arisen if they can be applied for tissue allografts sterilization. The selection of a sterilization dose is a compromise between a dose that is high enough to inactivate as many microorganisms as possible and low enough to preserve important biological properties of tissue allografts. In respect to human tissue allografts it is very difficult, or simply impossible to determine the bioburden each time, since initial contamination may vary greatly from tissue to tissue and from one donor to another. The problem is additionally complicated by the possible presence, in human tissues, of pathogenic viruses such as the human immunodeficiency virus (HIV), hepatitis viruses (HBV, HCV) or others. Data about the sensitivity of these viruses to ionizing radiation are scarce. The effectiveness of ionizing radiation to inactivate them in tissues that have been collected from cadaveric donors has not been well documented and the mechanisms of viral responses are unclear. The standards dealing with radiation sterilization do not take this into account because they concern health care products where biological aspects are not important. But irradiation has a dose-dependent effect on the plastic properties of bone and on the degree of collagen denaturation. Low temperatures during irradiation protect the osteo-inductive properties of allograft bones compared with irradiation at room temperature. And these factors have to be taken into account.

Dose of minimum 25 kGy is routinely applied in many tissue banks in the world. Using an in vitro model, with the samples at -70°C it shows that at least 25 kGy were required for complete viral inactivation. A dose of 25 kGy at -70°C is equivalent to a lower dose of irradiation at room temperature. It may therefore be possible to irradiate bone with a lower dose of electron beam irradiation at room temperature to achieve equivalent viral destruction. But it has also been be reported that at least 30 kGy were necessary for the destruction of the viral genome in bone-patellar tendon bone graft when the irradiation was performed with the grafts in dry ice. After many years of investigations the Central Tissue Bank in Warsaw has recommended dose of 35 kGy, so other multi-tissue banks in Poland have also implemented a dose of 35 kGy [5].

There are other problems associated with irradiation of tissue allografts which should be investigated to improve the quality of the sterilization process. For example the definition of batch and determination of batch size or the amount of samples that can be available for verification dose experiments. Amount of 100 pieces of allografts is far to big to perform the periodic audits. It is why the further investigations are needed [6].

4. Final Remarks

The dose setting procedures for medical devices sterilization are well established. The medical devices are manufactured in well known, stable conditions. It means that the bioburden (population of viable microorganisms on a product) can be easily determinate. The easiest case is when the sterilized product is homogenous, small in size, cheap and the sterilized batches are big and frequently radiation processing is applied. Petri dish is a good example of such a product. In this case the dose selection, product dose mapping and the sterilization dose auditing can be done easily. More complicated case is when the cost of product unit is high and production scale is small, then costs of performing sterilization dose auditing are big. But the worst situation is in the case of tissue allografts. There have not been written any standards concerning radiation sterilization of this type of materials. However, when radiation sterilization is applied to allografts it has to be aware about how to deal with the whole procedure of irradiation, starting from dose setting method at low temperatures and ending on sterilization dose auditing.

References

[1] EN 552: Sterilization of medical devices – Validation and routine control of sterilization by irradiation (CEN, European Committee for Standardization, Brussels, Belgium, 1994).

[2] ISO 11137: Sterilization of health care products- Requirements for validation and routine control – Radiation sterilization, (International Organization for Standardization, Geneva, Switzerland, 1995).

[3] EN 556: Sterilization of medical devices-Requirements for terminally sterilized devices to be labeled sterile (CEN, European Committee for Standardization, Brussels, Belgium, 1995).

[4] Kowalski J.B., Aoshuang Y., Tallentire A., Radiation sterilization - Evaluation of a new method for substantiation of 25 kGy, Radiat. Phys. Chem. 58, 77-86, (2000).

[5] Dziedzic –Goclawska A., The application of ionizing radiation to sterilize connective tissue allografs. In: G. O. Phillips et al. (Eds), Radiation Tissue Banking. World Scientific Publishing (In-Cooperation with IAEA), Singapore, London, NY, 35, 2000.

[6] Zimek Z., Kaluska I., Sterilization dose auditing for various types of medical products, Radiat. Phys. Chem. 63, 673-674,(2002).

Radiation Chemistry and Its Application to Radiation Technology

Vasil KOPRDA
Faculty of Chemical and Food Technology of Slovak Technical University, Radlinského 9, 812 37 Bratislava, Slovak Republic

Abstract. Ionising radiation interacting with matter produces the most different radiation-physical, radiation-chemical, and radiation-biological effects. In this contribution an accent on industrial applications of radiation sources is given. The possibilities of their application in chemical technology, namely in technology of radiation synthetic reactions, other polymeric reactions and to production of new macromolecular substances are reviewed. There are meant also radiation wastewater and sewage technologies, irradiation techniques in air cleaning processes, radiation sterilization in medicine, preparation of radiopharmaceuticals, etc. Some information on current possibilities of Slovak accelerators usage for irradiation technologies is brought.

Introduction

Ionising radiation in interaction with matter produces the most different effects. Some of them occur in nature, some only in laboratories and some also in industrial scale. Among the most significant industrial radiation applications belong:

- Radiation synthetic reaction,
- Industrial application of radiation for production of macromolecular substances,
- Radiation wastewater and sewage technologies,
- Radiation sterilizations in medicine and preparation of pharmaceutical products,
- Radiation technique in air cleaning processes,
- In food industry for enhanced preservation and quality improvement,
- In insecticide campaigns, radurization, radapterization, radicidation and disinfections.

In this contribution we try to cover important and very different effects of radiation that can be qualified as „new ways to solve old problems“.
From the scope of application of irradiation, it can be aimed to scientific problems, research and development, or to practical industrial applications in different chemical, technical, medical and biological fields, or in food industry and commerce.

In **biological fields** radiation namely used to microbiological applications, to liquidation of insects, bacteria, radiation modified fertility (for example tse-tse male fly sterilization in UNESCO Program for Africa, or propagating of "healing fly" etc.).

In **medical fields** at the first place sterilization of different materials comes into account, for example injection tubes and needles, ligatures, first aid kits, sterile gauze dressings, adhesive plasters, bandages, scalpels, scissors, forceps, and other chirurgical instrumentation, injection-ready solutions, surgical gloves, diagnostic paper strips, cotton pads, chirurgic tampons, cotton dressings, cotton swabs, different polymer foils, etc. Satisfied results were reached in radiation sterilization of Na salt of p-aminosalycilic acid, chloramphenicol, ampiciline, tetracycline, caffeine hydrochloride with 25 till 50 kGy. Using highly energetic radiation also NaCl isotonic solutions, glucose infusion solutions, atropine and morphine injection solutions, as well as medicaments in different vehicles and some vaccines were sterilized with doses 35 till 50 kGy.

Yet more extended field cover the production of radioisotope labeled diagnostics and **radioisotope labeled therapeutics**. These radio-pharmaceuticals contain in the function of radioactive tracer either reactor radioisotopes (emitters of beta$^{(-)}$ or gamma radiations – often used are ^{60}Co, ^{99m}Tc, ^{131}I and other – having longer half life and often emitting higher energy of radiation), that can serve as well for high dose tissue irradiation, or cyclotron-produced radioisotopes, pet-radioisotopes (usually positron or beta$^{(+)}$, short time and week or no gamma emitters – often ^{15}O, ^{13}N, ^{18}F, ^{11}C etc.) - more advantageous for investigated or cured patients. Positron emitter enables radically decrease the amount of single dose applied radioisotope. In the case of simple photon emission tomography around 1000 Bq of radioisotope is usually applied, whereas in the case of positron emission tomography 50 Bq is often adequate.

Among radiation applications in **food chemistry**, alimentary products and half-products industry two broad fields of interest are emerging:

- Antimicrobial attendance, sterilization of food half-products or ready-to-serve meals, quick meals, one-pot beverages like milk for 10o´clock (school) interval, or only packages sterilization,
- Pasteurization of alimentary products, mainly fish, oysters, and different "fruits of sea" (1-2 kGy), deeply frozen red meat (1.5-10 kGy), fresh fruit (0.3-1 kGy) and different spices.

In *technical fields,* ***radiation is namely applied to polymer chemistry (mixed organic polymers, natural-artificial mixed polymers, etc.), catalytic modifiers, heavy duty lubricants and others.***

1. Radiation Synthetic Reactions

Radiation induced reaction is of industrial significance only if it proceeds by chain mechanisms, the product is of great importance and can be produced in large scale amounts.

They can be well quantitatively characterized using term "radiation chemical yield G(X)" expressing an amount of radiation created product (of defined kind X) after absorption of 100 eV of radiation energy by irradiated matter. Radiation reactions belong to two groups:

- Reactions running with low G (units to tens units), with chemical yield till 10 % and are unique,
- Chain reactions with high G (from ten to hundreds and thousands), with chemical yield about 95%, that are far profitable compared to processes initiated chemically or with UV-light.

2. Industrial Applications. Irradiation Sources and Irradiation doses

Irradiation can be realized simply using **radionuclide sources** (for example, as in our University the massive sources of ^{60}Co isotope with half life 5.2 y and with maximum photon dose rate 0.5 Gy h^{-1}, or more often roentgen devices, and in the last years namely **particle accelerators** (electron beam, photon irradiation, light ion bombarding and heavy ions beam).

The radiation from technical irradiation sources used in radiation chemistry resembles the radiation from nucleus. Among the technical irradiation sources belong electron accelerators, which electrons reach energy comparable with energy of beta radiation of radioisotopes. Accelerated electrons are usually less scattered and have narrower energy distribution than negatrons. Disadvantage of an electron beam is its relatively small penetrability. In medium of unit density is penetrability about 5 mm per MeV. Accelerated electrons are limited to 6 MeV; higher energy starts activation processes in irradiated substances due to nuclear photo-effect and increases "bremstrahlung" (roentgen radiation) as well. The results of these limits are: the maximum irradiated thickness of unit density material is about 2 cm realizing irradiation from only one side, and 5 cm in the case of both sides. The basic problem of technological irradiation is a question of economy.

3. Changes of Features of Irradiated Substances

A proper dose of appropriate radiation can due to cross-linking change a thermoplastic substance convert into not fusible state of the substance, or decrease its solubility, or ability of water swelling, can change its form, appearance, mechanical features, electric conductibility, etc. Another effect can be initiated by gases, at the beginning diffusely dissipated and released from irradiated plastics after a temperature increase in fine bubbles, creating sponge plastics.

Technologically important and well investigated are effects of irradiation on mechanical and **optical features** of polyethylene (PE). PE-foil, common in wrapping technique is due to structural defects milky foggy. Being irradiated, the fog disappear, the foil become transparent and moreover it gain another feature - some shape memory. PE-foil after irradiation increases cross-bindings, and being stretched it keeps its dimensions, but after heating over 100 °C, the foil acquires the same shape as it had before.

Using high-energy radiation also natural resins can be **cross-polymerized**, caoutchouc can be vulcanized by that means and silicones as well. Another possibility is destruction of side chains or splitting the main chain in polymethylmethacrylate. In this reaction the most frequent products of **decomposition** are carbon dioxide (carbonate acid), carbon monoxide and hydrogen that at higher temperature create bubbles and change the polymer into sponge.

The well known is radiation polymerization and co-polymerization, for example acryl amide for extremely pure isolators and biomedical substitutes (Neutron Products Inc., USA).

The most interesting and promising application of radiation in polymer chemistry is possibility to combine more polymers together. Easy way is to **inoculation** of one kind of polymer onto another one. This technique was spread also on combinations of natural substances (cellulose, wood, etc.) with synthetic polymers. The impact of these techniques is very important for textile industry and also in electro technical industry. [1, 2, 3]

4. Accelerators in Slovakia

The first accelerators are being built in **Cyclotron Center of Slovak Republic (CC SR).** Their early use for cross-linking of selected thermo polymers is planned. New polymers will have increased thermal stability, ability of shrinking after heating, will show better mechanical properties (more resistant against cold flow, resistant against wear out, etc.). Among the different kind of application we can count also better isolation of links and cables, mechanical properties of tubes and lines, foils for package, better plastic spare parts, some new biomaterials, foam foils, supported building materials, that reach the same features with thinner walls. Further concern aims to electron beam curing processes using mixed polymers with fastened fibers for airplanes, navy and consumers applications.

Application of radiation in the field of chemical technologies is supposed to be distributed as follows: about 30 % for cross-linking of polymers, approximately 15 % for production of shrinking foils, about 15 % for enhancing elasticity and firmness of polymers, 15 % for improvement mechanical features of pipes and lines and about 15 % for medical sterilization. The rest covers the other minor industrial applications.

5. Workplaces of the Cyclotron Center of Slovak Republic

Project of accelerator technology applications in Slovakia counts to develop and run following workplaces of the CC SR:

- Center for positron tomography in Slovak Republic,
- Nuclear Medicine Center of the Central Military Hospital in Ružomberok,
- PET Center Bratislava,
- Mini PET-device Center,
- Laboratory of nanotechnologies and mesoscopic phenomena,
- Metrology auxiliary premises of the CC SR,
- Pavilion "J" – cyclotron DC 72,
- Therapy units of CC SR,
- Center of cold electron sterilization,
- Auxiliary premises of CC SR.

The Cyclotron Center within the Slovak Institute of Metrology (SIM) is being built since 1996. The Government of SR through its resolution No.407 (May 2001) decided to use CC SR also for industrial applications and joint-venture foundation. The building up of industrial CC SR workplaces with specialized aiming find itself in different stage of realization, (as a whole, can be said, in the second half of its realization). In the first time the workplaces are to be used namely in health service, further for education, for development of science and for industrial purposes. The project of CC SR would be finished till end of 2006.

The more detailed description of workplaces of CC SR gives better image on possibilities of their use in scientific, experimental or industrial scale. The state of realization of the first 10 workplaces (WP) around the end of 2003 is as follows:

WP1: The **workplace of the Positron Emission Tomography** (PET) of CC SR is situated at the Oncology Institute of Saint Elisabeth (OISE) in Bratislava. Clinical examinations are already provided with ^{18}F-deoxyglucose (FDG) on oncological and neurological patients. PET scanner ECAT EXAT HR^{+} from Siemens is used. This workplace is in full clinical operation. In OISE during the last three years about 1,600 patients were diagnosed. This Center is fully active also in educational and training courses. Results of PET examinations were presented at scientific events 16 times during the last year, 15 hours of lectures were devoted to post gradual education, 4 clinical trainings were organized and about 50 university students were educated or visited the Center.

WP2: This **workplace of CC SR at the Central Military Hospital in Ružomberok** is equipped with coincidence gamma scanner Millennium VG 5/8 System Hawkeye (General Electric Medical Systems, USA) Scanner enables fusion of images PET/CT and SPECT/CT and makes possible to examine the function of main body organs, motion set-up, inflammations and tumors, vascular system and others. Scanner serves basically for cardiology and provides services also for traumatology, neurology and other medical disciplines. Methods of nuclear medicine also makes possible to revert clinical problems with mapping of sympaticus on the heart by means of ^{123}I-MIBG (iodine traced metaiodobenzylguanidin) and ^{18}F-dopamin. Some hundreds of patients were examined there up today.

WP3: **PET Center in CC SR** is situated in pavilion I in the campus of SMI. Its subject of activity is production of short-lived radiopharmaceuticals (RAP) for region within the radius of 200 km (in Slovakia and neighboring countries). This Center is equipped with accelerator Cyclone 18/9 (IBA, Belgium), hot chambers (TEMA, Italy), gamma scanner PET/CT, Discovery and SPECT/CT, Millennium VG Hawkeye (GE, Med. Systems), Liquid chromatograph with mass spectrometric detector LC/MS Agilent 1100 (IAEA) for analysis of stability and control of radiopharmaceuticals (mainly ^{18}F-DG), Quality Control Laboratory for fluorodeoxyglucosis (LC with UVD and RMD, GC, Microsomometer (IAEA grant).

WP4. **Mini PET**, the small animal positron emission tomography for animals the size of mice and rats, designed to accurately quantify and visualize regional, time-varying distributions of proton-labeled RAP, namely:

- Investigate pharmacological, physiological, biochemical and genetic effects of new RAP,
- Characterize effects of RAP on various neurotransmission system after exposure to homotypic stressors or administration of different drugs,
- Concentrate especially on ^{18}F-derivatives of different neurotransmitters,
- Test RAP in small animals at different levels of homeostasis after acute or chronic stress exposure, during adaptation periods etc.,

in cooperation with Institute of Experimental Endocrinology of SAS, Center of Excellence of EU, IAEA, Vienna PET Clinical Center and NIH, Bethesda.

WP5. **Laboratory of Nanotechnologies and Mesoscopic Phenomena** is situated in pavilion "I" in the campus of SMI, Bratislava. This scientific, research and educational workplace is focused to development of nanotechnology methods and modifications of surface properties of solid substances and application of results in industrial practice. Also education and training program for students ob Universities in Bratislava will be provided under responsible authority of Faculty of Mathematics, Physics and Informatics of Comenius University.

WP6. **Metrology Auxiliary Premises in CC SR** of SMI. The planned activities will cover new fields such as metrology of nanotechnology (instruments, software, standards), metrology in chemistry, metrology in health care and molecular medicine (therapeutics), metrology of accelerators for applications, metrology ion PET, metrology in PET, metrology in ionizing radiation (proton and neutron therapy, electron beam sterilization), standardization of radionuclides for RAP, metrology for 3D structures.

WP7. Workplace **Isochronic Cyclotron DC-72**. This cyclotron, made in INR Dubna, RF, will be situated in pavilion "J", in campus SMI, Bratislava. Different beams will be available from DC 72: a proton beam (28 – 72 MeV), light ion beams from ^{2}H to ^{7}Li and heavy ions up to ^{129}Xe.

WP8. **Therapy Center** of CC SR The proton beam with maximum energy 72 MeV and reduced intensity below 50 nA will be used for proton eye therapy in pavilion "J" of SMI. For use in nuclear medicine production of therapeutic radionuclides ^{123}I and ^{211}At is planned. In the next, boron-neutron-capture therapy is expected.

WP9. The **Center of Cold Electron Sterilization** will be localized at Chirana T, Injecta Co., Stará Turá (West Slovakia factory for production of medical syringes and needles) and installation of electron accelerator from Saint Petersburg ought to take place to the end of 2004. Cold electron sterilization of medical products, of foodstuffs and some agricultural commodities is planned as well as some educational and training programs.

WP10. The last workplace of CC SR is **Center of Auxiliary Premises**, situated directly in objects of campus SMI. These promises (metal working machinery) will ensure the assembly and will be used for manufacturing of products based on new accelerator technologies (e.g. nucleopore filters). Small electron accelerator for use in metrology will be installed there.

6. Accelerators in Cyclotron Center SR

The CC SR is projected as a system established by two complementary accelerators:

1. **Base accelerator** of heavy ions and negative hydrogen ions
2. **Commercial low-energy accelerator** of negative hydrogen ions.

The base accelerator, isochronous cyclotron DC-72 produces ion beams from hydrogen (max. energy 72 MeV) to xenon (max. energy 2.7 MeV/nucleon). The maximum ion beam intensity is 6.10^{16} s^{-1} (100 μA) for $^{2}H^{-}$ and drops to value 6.10^{14} s^{-1} (1 μA) in the case of $^{129}Xe^{18+}$. The experimental devices placed in three extracted beam channels, three channels are for radiotherapeutical practice and two channels are designed for radionuclide production. Main radiochemical production of the DC-72 cyclotron is ^{123}I using a beam of 30 MeV negative hydrogen ions. Similar beam of lower energy (20 MeV) is for ^{81}Rb production. Hadrons therapy will exploit a DC-72 proton beam, 72 MeV for therapy of eye, for fast hadrons therapy and for boron neutron capture therapy. Original design source ECR, operating at 14 GHz frequency will be used as an injector of DC-72 cyclotron and also independently as a source of multi charged ions.

The commercial cyclotron unit for PET radionuclides fulfils the basic demand for commercial production 2 times 3 Ci (220 GBq) of ^{18}F within 6 (2x3) hrs run. Possibility to use deuterium negative ions accelerated to 6-7 MeV is also advantageous for some applications.

Among the important physical applications of energetic radiation belong:

- **Neutron** beams usage for data collection for hydrogen and helium production and for transmutation of radioactive wastes (Browman's-Rubio's concept),
- **Proton** beams utilization for measurement of cross sections of nuclides for monitoring reactions, for material study of defect formation, their spatial distribution and migration in radiation damaged materials, as well as study of effects on polymer-metal semiconductor interfaces,
- **Heavy ion** beams reaction mechanism studies of target bombarding, effects of conditions on it, fragments cooling with kinematic separator etc.

7. Program in Chemistry and Pharmacy

In the field of chemistry among the close future research activities belong the chemistry of resins, the development of adhesives, production of wood cotton by electron beam technology, production of viscose silk etc. Very interesting programs promises the use of electron beam in a field of biomaterials.

The hydro gel winded dressings (paintings, apertures etc.) were among the frontier commercial applications in this field. In the close future will follow:

- Micro gels and nano gels used as drug vehicles (transporter substances), often also with drug-release systems
- Functional polymers produced by splitting technique of vinyl monomers, especially with wet-sensitive features

- Network synthetic polymers and biologically important macromolecules.

Among the further planned applications of cyclotrons of CC SR belong:

- Bring into production already developed technology of polytetrafluorethylene (PTFE) nets formation using electron beam effects on melted substances
- Fibers production from silicone-carbide compounds using an irradiation of ante-ceramic polymers. Compounds are highly stretched and resistant against brittleness
- Develop new groups of amidooximes containing absorbents, developed using irradiation splitting of polyethylene fibers. They could help effectively collect the scarce metals (Uranium, Vanadium) from sea waters, or create effective filters for adsorption of environmentally harmful substances, e.g. Cadmium
- Recyclation of plastics and elastomers using electron beam technique.

The main objective of radiochemical and radiopharmaceutical program is cyclotron targetry, positron radionuclides and labeled compounds production for different applications. It should ensure quality assurance and control to provide a regular supply of high quality radiopharmaceuticals in a safe manner, and also perform input control of imported chemicals and radiochemicals (in the order of priority):

- $[^{18}F]$-FDG (fluorodeoxyglucose) and $[^{18}F]$-L-DOPA of radiochemical purity = 99.8 %
- $[^{123}I]$-NaI of radiochemical purity = 99.9999 %, for labeling of methyliodobenzylguanidine (MIBG), fatty acids, monoclonal antibodies and receptors
- $[^{11}C]$-CH_3 as a precursor for chemical synthesis (raclopride etc.), ^{11}C-methionin and other amino acids
- ^{81}Rb - ^{81}Kr radioisotope generators for ventilation scintigraphy
- $[^{13}N]$-NH_3 of radiochemical purity = 99.8 %
- $[^{15}O]$-H_2O or $[^{15}O]$-CO_2
- ^{211}At for research (monoclonal antibodies)

The usage of radiopharmaceuticals will be oriented towards:

- Oncological diagnostics (staging, recurrence, tumor viability, follow up etc.)
- Neurological diagnostics (tumors, epilepsy, dementia, Alzheimer and Parkinson disease)
- Cardiovascular indication (especially viability of myocardium).

About 2,000 patients will be examined annually once the CC SR is in full operation. Other extended applications are connected with radiotherapy and radiobiology. The most important applications are neutron capture therapy and proton eye therapy. Neutron capture therapy using boron compounds is used namely for brain tumor treatment. Possibility of irradiation with proton beam improves treatment of eye melanoma.

8. The Use of Radiobiological Effects of High Energy Radiation

Radiation attendance of food is physical method proper for:

- Prevention of microbial devaluation of food,
- Decrease of damages caused by insects and infections, as well as
- Ameliorate harms evoked by physiologic processes.

This method can substantially increase the store time of foodstuffs. In developing countries quantitative and qualitative losses of food are evaluated for 20 to 40%, at tropical conditions in some products 60 to 70%. Till today used methods - conservation, drying, deep freezing, thermal elaboration, vacuum packed products, etc. Insects, mushrooms, moulds and other offenders of agricultural products destroy annually according to data FAO about 33 million tons of wheat what represents production from area 4.8 million ha (large as Slovakia). Many countries practice irradiation of food and food products [4].

9. Radiation Technique in Water and Air Technology

The use of high-energy radiation in wastewater treatment and in the most other water technologies can not economically compete with traditional technologies [5]. A limited field of application exists in sewage technology. Irradiated and disinfected sewages, containing any toxic components, can be used to improve soil features (production of humine components), or as additives to fodder for fish and animals.

In air technologies can be high-energy radiation used for separation of air borne dust, oxides of Sulphur and Nitrogen. One of such technologies, developed in Japan is EBARA, where burn products of S and N are cooled, oxidized in electron radiation beam to sulphate and nitrate anions and next bound by chemical reaction with ammonia. Products can be used as fertilizers [6].

10. Radiation Insecticide Method

One of the most effective methods of high-energy irradiation sources exploitation seems to be radiation insecticide method. It is based on introduction of male insects, sterilized by radiation, into free environment. Radiation does not affect the pairing activity, but fertility. Once 95 % of males are sterilized, this species will die out in 8 generations, as females will produce only unfertilized ovums. Radiation insecticide method can be effective in many important cases in protection of humans (tse-tse fly, healing fly), in animals protection (cows after get rid of insects show an enhanced production of milk), or protection of fruit-production (Ceratis capitata etc.).

Acknowledgment

The author acknowledge the support by grant 2/2049/22 of VEGA Agency and kind support of ROSLO s.r.o. and REVING V.A.V. s.r.o., Bratislava.

References

[1] Griffin, G.J.F.: Chemistry and Technology of Biodegradable Polymers, Kluwer, Dodrecht, 1993.
[2] Ivanov, V.S.: Radiation Chemistry of Polymers (New Concepts in Polymer Sciences Series). Academic Press, New York, 1992.
[3] Krevelen van, D.W.: Properties of Polymers. Elsevier, Amsterdam, 1990.
[4] Tolgyessy J., Koprda V., Harangozó M., Tomeček O.: Priemyslové aplikácie elektrónových urýchľovačov, Acta Univ. Matth. Belii, Ser.Chem. No.6, 1-12, 2002.
[5] Tolgyessy J., Harangozó M.: Rádioekológia. FPV UMB, Banská Bystrica, 2000.
[6] Tolgyessy J., Dillinger P., Harangozó M.,: Jadrová chémia, FPV UMB, Banská Bystrica, 2000.

Physical, Chemical and Biological Dose Modifying Factors

Lajos G. GAZSÓ
National Center for Public Health, National Research Institute for Radiobiology and Radiohygiene
1775 Budapest, P.O. Box 101. Hungary

Abstract. The sensitivity of microorganisms towards high energy radiation varies widely: different types, species and strains exhibit greatly different radiation sensitivities. Certain environmental factors are also able to influence the actual radiation response. The basic principles of radiation damage, radiosensitivity of microorganisms and physical, chemical and biological dose modifying factors are reviewed in this paper.

1. Radiation Damage

The radiation induced inactivation of cells under a given test condition is a resultant effect of a series of complex physical, chemical and biological processes. Traditionally, it has been a practice to consider two quite distinct mechanisms. These have been called a direct and indirect actions [1]. The alteration in the molecule occurring as a result of absorption of radiation is said to be due to the direct action. The target may be ionized or excited initiating the chain events that leads to a biological change. On the other hand, when energy is absorbed in a certain molecule and transferred to a second molecule in which the chemical change takes place is called the indirect action (Figure 1).

This terminology is used successfully in studies on isolated cellular components, such as enzymes, nucleic acids. Application of this terminology for bacteria and fungi, however, has not been useful because of the chemical structural complexity of the cellular system.

Generally the damage to cells produced by ionizing radiation can be divided into three categories [2].

Lethal damage - which is irreversible, irrepairable, and by definition leads to cell death

Sublethal damage - which under normal circumstances can be repaired unless additional sublethal damage is added

Potentially lethal damage - this component of radiation damage can be influenced by environmental conditions (oxygen, temperature, chemicals, etc.)

Radiation action occurs over a broad timescale which extends from the very early physical processes to the very late biological effects, such as mutagenesis and carcinogenesis. The earliest event is the physical stage, which occurs between 10^{-18} – 10^{-12} second. The most important reactions of this stage are the ionization, excitation and dissociation of water, which lead to the formation of radical ions [3].

$$\begin{array}{ll} \text{Ionization} & H_2O \rightarrow H_2O^{+} + e^{-} \\ \text{Escitation} & H_2O \rightarrow H_2O^{*} \\ \text{Dissociation} & H_2O^{*} \rightarrow H^{\bullet} + {}^{\bullet}OH \end{array} \qquad (1)$$

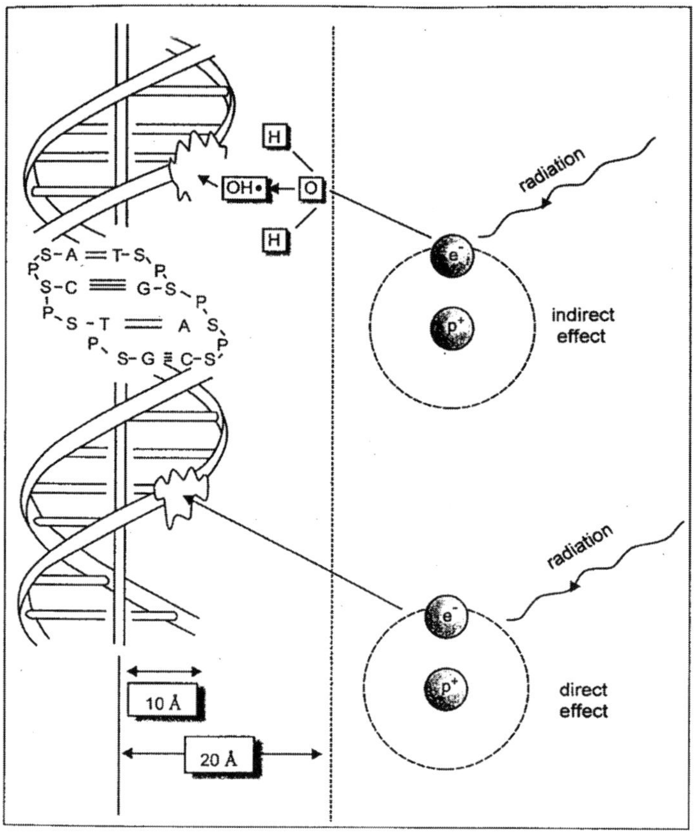

Figure 1. Direct and indirect effect of radiation.

The chemical stage occurs between 10^{-2} and 10^{3} second. The most important parts of it are the different reactions between primary products, homogenous distribution of free radicals and the biochemical processes.

The timescale of biological stage can range from hours up to several years (mutagenesis, carcinogenesis).

2. Relationships Between Dose and Effect

The relationships between dose and effect can be demonstrated by different kinds of survival curves. It is a common practice in the radiation biology to present results in the survival curves, where surviving fraction of organisms plotted semilogaritmically against dose of radiation. (Figure 2.) Originally three types of survival curve were described, namely exponential, sigmoidal and composite [4].

The exponential curve, a straight line when plotted as described above, it indicates that each organisms needs only one hit to be inactivated. The exponential curve can be fitted by the next equation,

$$S = e^{-kD} \tag{2}$$

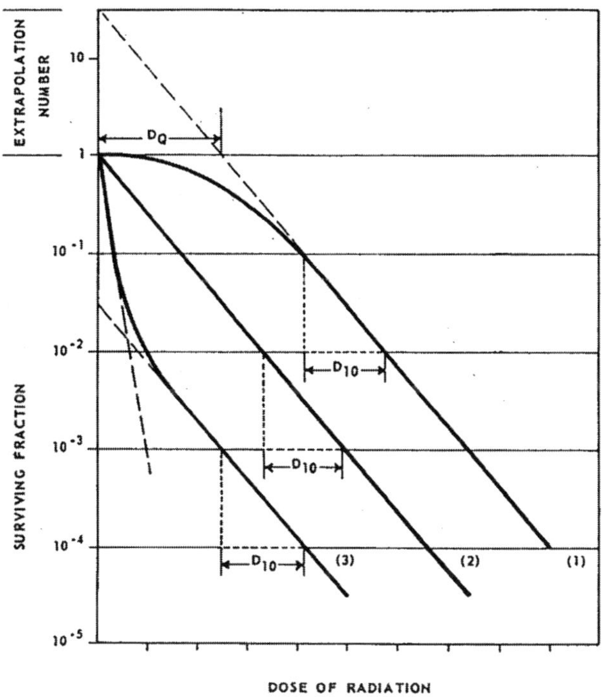

Figure 2. Hypothetical survival curves of irradiated bacteria: sigmoidal (1), exponential (2), and composite (3).

where S is the survival fraction after a single absorbed dose D, and k is the slope of the curve on semilogarithmic plot.

The sigmoidal curve is indicating that each organism needs more than one hit to be inactivated. This type of curve may be described by the so called multitarget single hit expression,

$$S = 1-(1-e^{-knD})^{n} \quad (3)$$

where the inactivation constant k is the sensitivity of each n target, each of which must be hit to kill.

In the case of composite curve, the population contains a mixture of two or more subpopulation (a sensitive and a resistant one) which separately would follow an exponential dose-effect curve. In the simplest case a mixture consisting of population a and b, the survival curve would be,

$$S = ae^{-kaD} + be^{-kbD} \quad (4)$$

For the practical application of radiation effect, the D_{10} value (decimal reducing dose) was introduced, which is the dose required to reduce the population by a factor of ten. The radiosensitivity of a microorganisms is conventionally expressed in term of D_{10} value. The unit of the absorbed dose is gray (Gy): 1 Gy = 1 J/kg[1].

3. The Radiosensitivity of Microorganisms

The sensitivity of microorganisms towards high energy radiation varies widely: different types, species and strains exhibit greatly different sensitivities. Certain environmental factors are also able to influence the radiation response (temperature, oxygen, water, soluble chemical agents).

Viruses

In general it is accepted that viruses are more resistant than bacterial spores. Single-stranded simple viruses are more sensitive than double-stranded complex structures [5]. Radiosensitivity of 30 viruses were studied by Sullivan at al. [6]. D_{10} values of viruses suspended in Eagle's minimum essential medium containing 2% fetal bovine serum ranged from 3.9 kGy to 5.3 kGy. The radiosensitivity was significantly affected by suspending media. The fully dry viruses are more resistant, as hydratation proceeds the radiosensitivity increases.

Bacteria

The bacteria show more complexity than viruses. From series of radiation microbiology studies, it can be concluded [7]:

- Among the vegetative bacteria, Gram-negative organisms (D_{10} ranging between 29 Gy - 240 Gy) are more radiosensitive than the Gram-positive species (D_{10} ranging between 180 Gy - 890 Gy).
- Bacterial spores are considerably more resistant than vegetative species. The anaerobic spore formers like Clostridium (D_{10} values ~ 2.2 – 5.4 kGy) are more radioresistant than aerobic Bacillus spores (D_{10} ranging between 1.2 and 5.0 kGy).
- Besides of the differences between the species, there are a number of factors concerned with the environmental conditions can greatly influence the actual radiosensitivity. For instance the D_{10} values of Salmonella typhimurium were significantly different, when the suspending medium was phosphate buffer (D_{10} = 210 Gy) or fish meat, where D_{10} value reached the 1.74 kGy [8]. The different supporting surfaces can also alter the radiosensitivity of bacteria.
- The bacterium Micrococcus radiodurans isolated from irradiated meat is the most radiation-resistant organism known. D_{10} values can reach 10 kGy. The radiation resistance of this strain has been attributed to its exceptional repair capabilities rather than to an altered susceptibility to radiation of its genetic material per se [9]. Thus, the ability to repair DNA double-strand breaks has been reported. The specific nature of this repair is still not clear, though it is certain that M. radiodurans possesses DNA excision repair and DNA recombination activities. The taxonomy study has been suggested that Micrococcus radiodurans and its relatives (M. radiophylus, M. radioproteolyticus) are distinct from conventional Micrococcus species [10]. Structural observations on these organisms emphasize the unique features of their cell wall and membranes. A new name was introduced as Deinococcus radiodurans.

Fungi

Most studies of the inactivation of fungi by irradiation have been made on asexual spores. Germinating spores, mycelia and other morphological structures of fungi might have different radiation responses [11]. The radiation sensitivity of fungi is influenced not only by genetic factors but also by the number of cells in a spore (effect of multicellularity), the number of nuclei per cell (effect of multinuclearity). The haploid yeast cells are more sensitive than diploid ones (effect of ploidity). Ten species of fungi representing the genera Alternaria, Aspergillus, Cladosporium, Curvularia, Fusarium and Penicillium were examined by Saleh et al., [12]. D_{10} values of fungal conidia in water for Aspergillus niger 420 Gy, for Cladosporium cladosporoides 300 Gy and for Curvularia geniculata 290 Gy.

D_{10} values for dematiaceous fungi (in agar medium!) ranged from 6 to 17 kGy and for moniliaceous fungi were less than 3 kGy. Yeast appear to be about as sensitive as non-sporeforming bacteria.

At present an immense quantity of data is available in the literature. Unfortunately, most of these data were obtained under different experimental conditions. Regarding the considerable influencing effect of environmental condition to the actual radiosensitivity, to achieve a correct comparison is very difficult.

4. Dose Modifying Factors

The radiosensitivity of microorganisms can be influenced by some factors other than genotype of species. The responses of cells to a given dose can be altered in different ways. This is possible because response depens on physical factors (quality of radiation, temperature, etc.), on chemical factors (oxygen, water content, chemical agents, etc.) and the biological or physiological factors (growth phase, amount of DNA). (Table 1.)

The dose modification can be expressed by the dose modification ratio (M) - this is the ratio of dose under reference conditions to test conditions to produce the same level of effect.

$$M = D_R/D_T \tag{5}$$

4.1. Physical Dose Modifying Factors

The radiation damage is highly depending on the *quality of radiation*. The radiation quality can be caracterized by the Linear Energy Transfer. LET is defined as the energy lost by particle per unit length of medium. To describe the difference between high LET (fast neutrons, accelerated heavy ions, etc.) and low LET (γ-ray, X-ray, etc.) radiation, the Relative Biological Effectiveness was introduced. RBE is a ratio of absorbed dose of a reference irradiation (D_R) to the absorbed dose of test radiation (D_T) to produce the same level of biological effect,

$$RBE = D_R/D_T \tag{6}$$

The value of RBE depends on the radiation dose, the dose rate and the biological system. The relationship between LET, RBE and OER can be seen on Figure 3.

The radiobiological importance of high LET particles:

- relative biological effectiveness is increased
- oxygen enhancement ratio is reduced

Table 1. Dose modifying factors.

Phychical	Chemical	Biological
Quality of radiation	Oxygen	Amount of DNA
Dose rate	Water content	Growth phase
Dose fractionation	Sensitizers	Cell cycle
Temperature	Protectors	Age
	$^{\bullet}$OH-scavengers	Sex
	Antioxidants	
	Thiol-reactive agents	

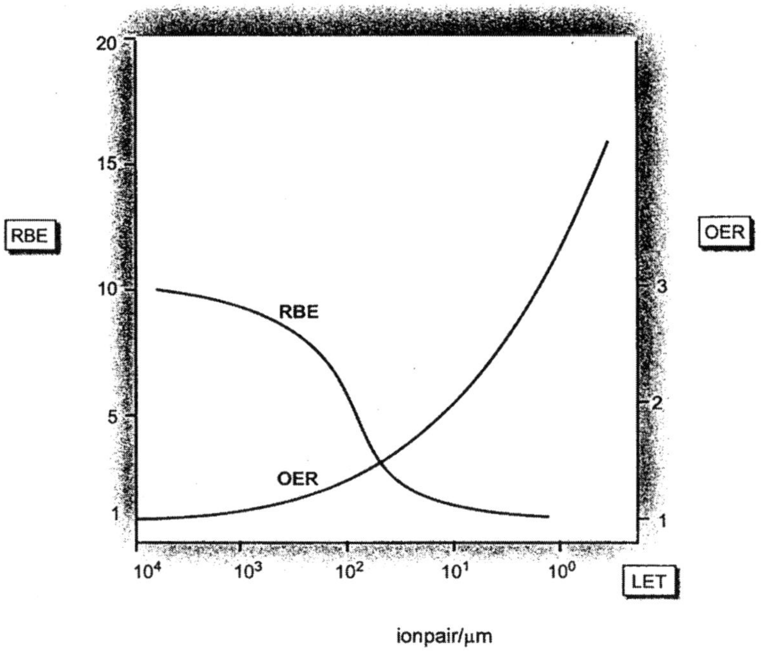

Figure 3. Relationship between LET, RBE and OER.

– repair of radiation damage is less
– the age response function is suppressed.

Dose rate differences between γ-rays are too small to be of any significance with respect to microbial inactivation, but clearly the large difference between ^{60}Co gamma-ray and accelerated electrons. High dose rate may decrease the efficiency due to the radiolytic depletion of oxygen [13].

Very soon after X-ray was began to be used clinically it was recognized that the radiation response was usually reduced if the total dose was delivered in fraction rather than single shoot. In the case of *dose fractionation* a second shoulder appears in the surviving curve. The manifestation of a second or more shoulders assumed to be evidence that radiation damage must accumulate within the cell before a final event becomes lethal. This sublethal damage during the so called "recovery interval" can be restored. The size of the shoulder depends on the repair capacity of cells and on the recovery interval, which usually ranged from minutes up to hours.

The *temperature* is also an important physical dose modifying factor. Experiments with dry spores of Bacillus megaterium shows that the sensitivity is constant between -268°C and -148°C, increasing temperature up to 20°C results an increased sensitivity by about 40%. The influence of temperature is similar for oxic and anoxic spores [14]. Fully hydrated Bacillus megaterium spores equilibrated with oxygen the sensitivity increases by 16% on decreasing temperature from +18 to + 2.5°C. Further reduction in temperature down to -196°C decreases the level of radiosensitivity by about 45%. In contrast, for anoxic spores, the radiosensitivity increases sligthly with decreasing temperature from +18°C to + 5°C. Reduction in temperature to -196°C results only a small decrease in sensitivity [15].

A number of investigations have reported that relatively mild doses of ionizing radiation sensitized bacterial spores (and many other microorganisms as well as viruses) very significantly to subsequent heat [16]. Combined heat and radiation treatment of microorganisms yields a lethal effect greater than the additive rates of independent agents [17]. Maximum synergism occurs at those conditions where heat and radiation are equally effective as destruction agents.

4.2. Chemical Dose Modifying Factors

Oxygen has received the greatest attention of all chemical agent known to modify radiation damage in cells. Oxygen has been found to increase the sensitivity to radiation of almost all type of cellular systems and even higher organisms, and this phenomenon has been generally known as the "*oxygen effect*".

The sensitizing effect of oxygen can be expressed by the Oxygen Enhancement Ratio (OER), which is the ratio of dose required under anoxic condition to that under condition of air to produce the same level of effect.

Gray [18] considered at first the possibility that the action of oxygen is mediated through reaction with product of the radiolysis of water. These reactions can be presented in a simplified form as follows,

$$\begin{aligned} &H_2O \rightarrow HO^{\bullet} + H^{\bullet} \\ &H^{\bullet} + O_2 \rightarrow HO_2^{\bullet} \\ &e_{aq}^{-} + O_2 \rightarrow O_2^{\bullet -} \end{aligned} \tag{7}$$

Secondary $HO^{\bullet}$ radicals can be produced from superoxide anion and hydrogen peroxide through the Haber-Weiss reaction [19].

Later Howard-Flanders proposed that the irradiation creates two types of damage [2],

$$\begin{aligned} &R \xrightarrow{radiation} R^{''}(lethal) \\ &R \xrightarrow{radiation} R^{'}(potentially\ lethal) \\ &R^{'} + O_2 \rightarrow R^{''}O_2(lethal) \end{aligned} \tag{8}$$

The potentially lethal damage is not lethal to the cells unless it reacts with oxygen. Later three distinct responses obtained by altering the gaseous environment [20, 21]. The authors designated three kinds of damage, (Figure 4.)

Class I. damage, is seen when the cells are in anoxic condition during and after the irradiation. This damage is independent of oxygen.

Class II. damage, oxygen dependent and it is called as "immediate oxygen dependent damage". It occurs when oxygen is present during and after irradiation. It is believed to result from interaction of oxygen with short-lived radicals.

Class III. damage, is the post irradiation oxygen dependent damage, which occurs when the cells are in anoxic condition during irradiation and oxic condition after the irradiation. This damage is known to occur as the result of interaction of oxygen with long-lived free radicals, formed by irradiation in absence of oxygen.

The oxygen dependent sensitization is quite similar in the cellular radiobiology. The OER values are varied between 2-4.

For the radiation chemical mechanisms of oxygen effect two hypothesis have been proposed [22]. Namely, the "oxygen fixation" and the "activated oxygen" hypothesis.

The *chemical radiosensitizing agents* have practical value in the treatment of cancer with radiation. Regarding to the whole natural environment, many chemicals can enhance the radiation response. Within the last decades, various classes of chemical agents have been found to increase the efficiency of radiation induced cellular damage. These include inorganic and organic chemicals with various properties.

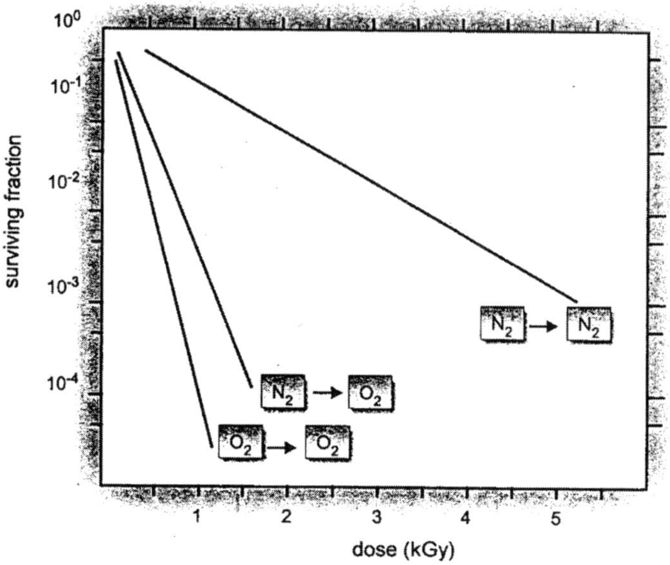

Figure 4. Surviving curves of bacterial spores irradiated under different gaseous conditions.

a) Sensitizers specific for hypoxic cells
   - electron affinic agents
   - membrane-specific agents
b) Analoguos of DNA precursors
   - incorporated into DNA
   - non-incorporated into DNA
c) Radiation-activation cytotoxic compounds
d) Agents which modify cellular regulatory process
   - inhibitors of repair
   - DNA-binding and intercalating compounds
   - inhibitors of natural radioprotectors

From practical point of view the electron affinic sensitizers and inhibitors of natural radioprotectors play an inportant role in the cellular radiobiology. A large number of radiation sensitizing compounds of "electron affinic" class have been developed and tested in vitro and in vivo [23]. The electron affinic agents are good scavengers of hydrated electrons when increases the yield of OH radicals, such a reaction between N_2O and e^-_{aq}

$$N_2O + e^-_{aq} + H_2O \rightarrow N_2 + OH^- + HO^\bullet \tag{9}$$

Scavenging the hydrated electron into OH radicals by electron affinic sensitizers prevent the reaction,

$$e^-_{aq} + HO^\bullet \rightarrow OH^- \tag{10}$$

which in absence of sensitizers converts the radical into harmless OH^-.

Natural thiols, mainly represented by glutathione, can also influence the radiation sensitivity [24]. Glutathione can modify the radiation induced damage by scavenging radicals of the radiolysis of water and by hydrogen transfer to target radical. Glutathione may also involved in enzymatic repair processes by glutathione reductase, glutathione peroxidase and number of thiol-disulfide exchange enzymes [25]. Inhibitors of natural radioprotectors, namely thiol reactive agents can decrease the actual glutathion content enhancing the radiation response [26].

Protective agents are chemicals, which reduce the lethal effect of radiation. The most remarkable group of protectors are the sulfhydryl compounds, which were discovered many years ago [27]. Agents such as cysteine, mercapto-ethyl-amine and amino-ethyl-isothiuronium were among the most effective. Favoured hypothesises are the hydrogen donation from the -SH (as a reaction competing with damage fixation) and the ability to quench free radicals and their products [28].

The water content of the microbial cell at the instant of irradiation is also known to affect greatly its radiation response. For spores in N_2, progression from the wet to dry state causes a lessening in response of radiation, whereas for oxygenated spores, a similar progression results an increase in response [29]. Similar overall water effect have been recognized in vegetative bacteria and mould spores.

4.3. Biological Dose Modifying Factors

Effect of radiation on cells can be modified not only agents present during irradiation, but also biochemical processes occuring over a much longer time. Profound changes in radiation response may be altered to progress of cell throught different phases and stages of growth. These may be associated with changes in the intercellular environment. The literature on the radiation response of bacteria in different growth phases reveals some contradictory results [30]. Unfortunately a general rule concerning the influencing effect of growth phase is not available. Sometimes survival curves are deeper in exponential than in stationary phases, sometimes the reverse is true. Shoulder of curves may be seen when microorganisms are in stationary phase, but not when the cells are growing exponentially or vice versa. Differences may be attributed to individual biological nature of the strains used.

Some data are available concerning the effect of postirradiation cultivation conditions. The medium can influence the post-irradiation recovery. Alper and Gillies [31] reported that suboptimal growth conditions were best.

In this paper we tried to describe the extent of the radiosensitivity of microorganisms and factors, other than genotye of microbes influencing radiosensitivity.

References

[1] Hall, E.J. (1978) *Radiobiology for Radiologist*, Harper and Row Publisher Inc., New York.

[2] Howard-Flanders, P. (1958) Physical and chemical mechanisms in injuring of cells by ionizing radiations, Adv. Biol. Med. Phys., 6, 553-558.

[3] Tubiana, M., Dutriex, J. and Wambersie, A. (1990) Introduction to Radiobiology, Taylor and Francis, London, New York, Philadelphia.

[4] Gunter, S.E. and Kohn, H.I. (1956) The effects of X-rays on the survival of bacteria and yeast, I. Bacterial, 71, 422-428.

[5] Pollard, R.C. (1973) *The effect of ionizing radiation on viruses,* in Manual on Radiation Sterilization of Medical and Biological Materials, IAEA, Vienna, 61-65.

[6] Sullivan, R., Fassolitis, A.C., Larkin, E.P., Read, R.B. and Peeler, J.T. (1971) Inactivation of thirty viruses by gamma radiation, *Appl. Microbiology,* 22, 61-65.

[7] Sztanyik, L.B. (1974) *Application of ionizing radiation to sterilization,* in E.R.L. Gaughran and Goudie (eds), Sterilization by Ionizing Radiation, Multiscience Publication Limited, Montreal, 6-38.

[8] Ley. F.J. (1973) *The effect of ionizing radiation on bacteria,* in Manual on Radiation Sterilization of Medical and Biological Materials, IAEA, Vienna, 37-63.

[9] Hansen. M.T. (1978) Multiplicity of Genome Equivalents in the Radiation-Resistant Bacterium Micrococcus radiodurans, *Journal of Bacterology,* Vol. 134, No.1. 71-75.

[10] Brooks, B.W. and Murray, R.G.E. (1981) Nomenclature for „*Micrococcus radiodurans*" and Other Radiation-Resistant Cocci: *Deinococcaceae* fam. nov. and *Dienococcus* gen.nov., Including Five Species, *International Journall of Systematic Bacterology, Vol.31. No.3. .353-360.*

[11] Sommer, N. (1973) *The effect of ionizing radiation on fungi*, in Manual on Radiation Sterilization of Medical and Biological Materials, IAEA, Vienna, 73-79.

[12] Saleh, Y.G., Mayo, M.S. and Ahearn, D.G. (1988) Resistance of some fungi to gamma irradiation, *Appl. Environmental Microbiology,* 54, 2134-2135.

[13] Adams, G.E. and Stratford, I.J. (1978) *Some dose rate effect in irradiated microorganisms,* in E.R.L. Gaughran and A.J. Goudie (eds), Sterilization by Ionizing Radiation, Multiscience Publication Limited, Montreal, 81-96.

[14] Powers, E.L., Webb, R.B. and Ehret, C.F. (1959) An oxygen effect in dry bacterial spores and its temperature dependence, *Exptl. Cell. Res.,* 17, 550-557.

[15] Gazsó, L.G. and Tallentire, A. (1981) The influence of temperature and phase state on the radiosensitivity of Bacillus megaterium spores (in Hungarian), *Izotóptechnika,* 24, 27-31.

[16] Sivinski, H.D. (1975) *Treatment a sewage sludge with combinations of heat and ionizing radiation,* in Radiation for a Clean Environment, IAEA, Vienna, 151-167.

[17] Fisher, D.A. and Pflug (1977) Effects of combined heat and radiation on microbial destruction, *Appl. Environmental Microbiology,* 33, 1170-1176.

[18] Gray, L.H. (1961) *Mechanisms involved in the initiation of microbiological damage in aerobic and anaerobic system,* in R.J.C. Harris (ed) The Initial Effects of Ionizing Radiation on Cells, Academic Press, New York, 21-44.

[19] Haber, F. and Weiss, J. (1934) Catalytic decomposition of hydrogene peroxide by iron salts, *Proc. R. Soc. London,* (Ser.A) 147, 332-351.

[20] Ewing, D. and Powers, E.L. (1976) Irradiation of bacterial spores in water: three classes of oxygen-dependent damage, *Science,* 194, 1094-1096.

[21] Ewing, D. and Powers, E.L. (1980) *Oxygen-dependent sensitization of irradiated cells,* in R.E. Meyn and H.R. Withers (eds) Radiation Biology in Cancer Research, Raven Press, New York, 143-168.

[22] Greenstock, C.L. (1984) Oxy-radicals and the radiobiological oxygen effect, *Israel J. Chemistry*, 24, 1-10.

[23] Adams, G.E. Clarke, E.D., Flockhart, I.R., Jacobs, R.S., Sehmi, D.S., Stratford, I.J., Wardman, P. and Watts, M.E. (1979) Structure activity relationship in development of hypoxic cell radiosensitizers, *Int.J. Radiation Biology,* 35, 133-150.

[24] Kosower, N.S., Kosower, E.M., Newton, G.L. and Ranney, H.L. (1978) Glutathion status of cells, *Int. Review Cytology,* 54, 109-160.

[25] Edgren, M., Révész, L. and Larsson, A. (1981) Induction and repair of single-strand DNA breaks after X-irradiation of human fibroblast deficient in glutathione, *Int. J. Radiation Biology,* 40, 355-363.

[26] Gazsó, L.G. and Dám, A.M. (1990) *Stabilization of enhanced radiosensitivity of Bacillus megaterium spores by pretreatment of di-ethyl-maleate and diamide,* in E. Riklis (ed) Frontiers in Radiation Biology, Balaban Publishers, Weinheim, 229-234.

[27] Akerfeldt, S. (1963) Radioprotective effect of S-phosphorylated thiols, *Acta Radiol.,* 1, 465-471.

[28] Yuhas, J.M. (1983) *Pharmacokinetics and mechanisms of action of WR-2721 and other radioprotective agents,* in O.F. Nygaard and M.G. Simic (eds) Radioprotectors and Anticarcinogenesis, Academic Press, New York, 639-653.

[29] Tallentire, A. and Powers, E.L. (1963) Modification of sensitivity to X-irradiation by water in Bacillus megaterium, *Radiation Res.,* 20, 270-287.

[30] Alper, T. (1980) Cellular Radiobiology, Cambridge University Press, Cambridge.

[31] Alper, T. and Gillies, N.E. (1958) Restoration of Escherichia coli strain B after irradiation its dependence on suboptimal growth condition, *J. Gen. Microbiology,* 18, 461-472.

Radiation Inactivation of Bioterrorism Agents
L.G. Gazsó and C.C. Ponta (Eds.)
IOS Press, 2005

Radiation Technology in the Mediterranean Dialogue Countries

Yousri RAAFAT
National Center for Radiation Research and Technology (NCRRT), Atomic Energy Authority, P. O. Box 29, Nasr City, Cairo, Egypt

Abstract. Egypt has already an advanced radiation processing programs in different areas such as medical sterilization, food irradiation and industrial applications. Such technologies already are existed in Egypt since 1972 at that time The National Center for Radiation Research and Technology (NCRRT) was established. It was aiming at promoting research and development using ionizing radiation. Since that time, Egypt has been engaged in collaboration with Mediterranian countries through IAEA, AFRA member states and Arab Atomic Energy Agency. Radiation processing of medical and food products in addition to industrial applications is the main concern and consideration at NCRRT. The center is divided into three divisions, including twelve departments, in addition to central laboratories and industrial radiation processing facilities. The main objectives of NCRRT are to facilitate the concrete application of radiation technology in environmental studies as well as the processing of selected materials for the benefits of peaceful applications in our daily life.

Introduction

The main objectives of NCRRT are to facilitate the concrete application of radiation technology in environmental studies and industrial applications. In the following the main activities are summarized:

- Extension of the market capacity for radiation-processed products in order to generate a substantial income is necessary. Therefore, a new gamma irradiation facility will be installed at Alexandria City (near the seaport) for the export of sterilized medical products as well as irradiated fresh and dried foods. The first stage of the project is completed, while the second and the final stages will be completed within the coming two years.
- Intensify the collaboration with the industrial companies, such as CID Pharmaceutical Co., to finalize the proper implementation of EB accelerator for the commercial production of wound-dressing hydrogels. A contract was issued and signed with the Ministry of Health for that purpose.
- Discussion is in progress with VIVIRAD Co. (formerly, High Voltage Corporation) to upgrade the existing electron accelerator, specifically to increase the beam energy from 1.5 MeV to 3 MeV. Such an upgrade requires a funding of about 0.5 Million US$. We are hoping that this can be supported through a bilateral agreement between Egypt and the IAEA. Through this modification, the following can be accomplished :
  - Production of wound-dressing hydrogels of different thickness.
  - Production of heat-shrinkable materials by radiation crosslinking technology.

- Crosslinking of larger wire and cable insulators.
- Recycling of agriculture waste products.
- Production facility for radiation-sensitive indicators is being expanded to meet the demand. The R&D for these indicators was carried out at the NCRRT. They are used for process control for radiation applications. We are aiming to fulfil the needs of the region.
- Improvement of the physical and mechanical properties of wires and cables of different sizes. Several meeting have been held with a large commercial company (El-Seweedy Industries) in this regard. They have visited European facilities to investigate feasibility and now are quite keen on proceeding with NCRRT. However, higher electron beam energy would be necessary to proceed with such industrial applications.

- The enhancement of the utilization of the peaceful applications of the nuclear energy to the welfare and sustainable development of Egypt.
- The adoption of the principle of full transparency regarding all nuclear activities in Egypt.
- The commitment to internationally accepted nuclear treaties promoting the peaceful applications of nuclear energy (for example the NPT) and compliance to the international safeguard system for unclear materials as long as these commitments don't endanger the national security of Egypt.
- Securing the radiological environmental safety of Egypt, from both internal and external radiation hazards.
- Enhancing the public awareness of the benefits of the peaceful uses of nuclear energy and addressing the public concern on nuclear issues.

Technical Status and Achievements at NCRRT

1. Gamma Facility and Laboratory Sources

Mega Gamma 1 type J-6600 irradiator is a self-automated industrial Co-60 facility and is applied for radiation sterilization of medical products and devices and foodstuffs irradiation. It was constructed in 1979 with an activity of 500 kCi. The current activity is close to 275 kCi. The recently upgraded transport and control system from analog to digital using PLC system was done in year 2000 and it allows perform the automatic operation at certain irradiation dose level and screening all the information related to the facility. Safety conditions and ventilation equipment are provided according to IAEA instructions. Suitable loading and unloading zones are present to fulfill necessary storage surfaces in the facility. The gamma plant has a tote box conveyor system of 0.25 cum per tote box.

Five laboratory size gamma sources are used for basic and apply research. They are as follow:

- Gamma Cell model GC 220, MDS Nordion, Canada. Activity Co-60 sources at March 2001 : 91 Ci, dose rate 0.019 Gy/s,
- Cesium Gamma Cell model GC 40, MDS Nordion, Canada. Activity Cs-137 sources at March 2001:91 Ci, dose rate 0.009 Gy/s.
- Indian Gamma Cell model 4000A, Bah Bah Research, India. Activity Co-60 sources at March 2001:1897 Ci, dose rates 1.4 kGy/h.

- Russian Gamma Cell model Issledovatel, Russian Atomic Energy Authority, Russia. Activity Co-60 sources at March 2001:10642 Ci, dose rate 7.1 kGy/s,
- Gamma Cell model 220, MDS Nordion, Canada. Activity Co-60 sources at March 2002: 10 kCi.

The gamma laboratory sources installed at NCRRT are sufficient for wide research program implementation and high –dose reference dosimetry laboratory use.

2. Electron Beam Accelerator Facility

The electron accelerator was installed in the Center under IAEA technical and financial support. Accelerator is installed vertically in concrete shelter placed in technological hall. The nominal technical specification of the accelerator is as follow:

- Electron energy 1.5 MeV,
- Beam current up to 25 mA,
- Beam power up to 37.5 kW,
- The length of the scanned beam spot 90 cm.

The other instruments provided by IAEA under technical support program for accelerator facility are as follow:

- Conveyor system,
- Transport system for continuous irradiation of wires and tubes,
- Chart recorder,
- Set of test and maintenance tools and accelerator spare parts.

3. Radiation Safety Systems and Procedures

56 persons in NCRRT were considered as radiation workers due to their activity related to exploitation of sources of ionizing radiation (gamma facility; 5 gamma sources; electron accelerator) in 2002. The basic radiation protection measures at the NCRRT are mainly related to International Basic Standard, IAEA, No 115, 1994 and International Commission of Radiological Protection, 1991. The practical application of these measures are dealt with three principles:

- no occupational radiation exposure should be adopted unless it produces sufficient benefit,
- personnel doses should be kept as low as reasonably achievable (ALARA) taking into account economic and social factors,
- the individual doses should be subjected to specified limit: whole body dose equivalent of 20 mSv/y for occupational exposure and 1 mSv/y for public, considering work load as 2000 h/y.

The following radiation monitoring instruments are used to control external radiation exposure around gamma sources installed at NCRRT:

1. Eberline X-gamma monitor up to 30 mSv/h,
2. ND-3000 X-gamma monitor up to 10 mSv/h,
3. Ludlum model 44-88 detector for alpha and gamma radiation.

The maximum gamma radiation levels around NCRRT irradiation facilities were carefully investigated at different positions (top, around shield, 1 m distance, control panel).

It should be noticed that dose rates at some positions around the irradiation facilities are higher than radiation protection limits. These positions are very occasionally occupancy areas during maintenance of this unit under the radiation protection supervision.

4. High-Dose Reference Dosimetry Laboratory

A reference laboratory for high-dose dosimetry has been established at NCRRT to provide traceability for the calibration of dosimetry systems required for radiation processing facilities with electron accelerators and gamma sources. The main activities of High-dose Reference Dosimetry Laboratory are related to:

- traceability to industrial and research facilities,
- dose mapping procedures,
- training programs on dosimetry and quality assurance,
- inter-comparison dose programs,
- immediate assistance in case of problems arise at an industrial gamma and electron facility,
- provides seminars for personnel at their facilities on information related to dosimetry and process control

4.1. Reference Dosimetry Systems are Based on

- Dichromate aqueous system for gamma rays,
- Polystyrene (PS) colorimeter for electron beam,
- Alanine-EPR measurements both for gamma and electron beam,
- Ethanol chloro-benzene (ECB) dosimetry system.

4.2. The Principal Equipment Located in High-Dose Reference Dosimetry Laboratory is as Follow

- Unicam UV/VIS spectrometer model 8625,
- Polystyrene calorimeter manufactured by Riso Reference laboratory,
- Dedicated ESR spectrometer for alanine dosimeters measurements,
- Environmental control chamber model 518,
- Millipore system of water purification,
- Co-60 source Gamma Cell 220-2.

5. Equipment at Central Labs

The NCRRT laboratories are equipped with high number unique analytical instruments. The instruments listed bellow are used in different scientific applications and investigations on selected processed materials:

- X-rays Diffraction (XDR) MODEL DP-DI, Shimadzu, Japan,
- X-rays Fluorescence Analyzer model EDX4, Philips,
- X-rays Forces model DX-95,
- Inductively Coupled Plasma-Atomic Emission Spectrometer (ICP-AEA) model Ultima, Joban Y von, France,
- Nuclear Magnetic Resonance (NMR) model NMR-300 MHz, Bruker, Germany,

- Electron Spin Resonance (ESR) model EMX, Bruker, Germany,
- Ultramicrotome model III, Co. LK13, for sample preparation for transmission electron microscope,
- Transmission Electron Microscope (TEM) model JEM-100 CX, JOEL, Japan,
- Scanning Electron Microscope (SEM) model JEOL-JSM 5400, JEOL Japan,
- High Performance Liquid Chromatography (HCLP) model WIS 201 + FL 2000 Detectors, Thermo Separation Products, USA,
- Amino Acid Analyzer (AAA) model Biochrom 20 Swede, Pharmacia Biotech,
- Atomic Absorption Spectrophotometer (AAS) model Unicam 939, Unicam, England,
- High Performance Microwave Digestion System model MLS 1200 ega.
- Differential Thermal Analysis (DTA) model DTA-50, Shimadzu, Japan,
- Differential Scanning Calorimeter (DSC) model DSC-50, Shimadzu, Japan,
- Thermal Gravitational Analysis (TGA) model TGA-50, Shimadzu, Japan,
- Thermal Mechanical Analysis (TMA) model TMA-50, Shimadzu, Japan,
- Gas Chromatography/Mass Spectrometer (GC-MS) model Finnigan SSQ 710, Finnigan Mat, USA,
- Gas Chromatography/Mass Spectrometer (GC-MS) model hp 890, Hewlett Packard, USA,
- Super Critical Fluid Extraction (SFE) model hp 7680 T, Hewlett Packard, USA.

6. Major Field of Interest in the Field of Radiation Technology at NCRRT

- Radiation processing of food that includes microbiology, nutrition and biochemistry aspects of irradiated foods and animal feed.
- Electron beam processing of polymers, biomaterials and surface curing of materials.
- Radiation sterilization of healthcare products.

6.1. Radiation Processing of Food is one of the main activities of NCRRT and it was started in1973. The Department of Food Irradiation has several laboratories namely food technology, biochemistry, microbiology, nutrition, wholesomeness and radiation dosimetry. There are more than 200 Ph.D. and M.Sc. thesis produced by NCRRT and the universities in relation to food irradiation and 6 thesis in relation to the economic aspects of the technology.

In June 1996, the Egypt Health Authority has granted an unconditional clearance for food irradiation, up to the level of 10 kGy for herbs, spices and dried onion and dried garlic. This is a commitment of the Government and NCRRT in particular to the application of gamma radiation in reducing losses and contamination due to microbial pathogens, bacteria, yeast, moulds as well as insects and parasites. Other food items such as fresh fruits are still under consideration of the Ministry of Health in collaboration with NCRRT and Ministry of Agriculture. Irradiation services for food are provided by Mega Gamma irradiation facility at NCRRT, Cairo. It was estimated that~50 tons of food items per annum are irradiated at Mega Gamma irradiation facility.

Food Products Treated by Gamma Rays

Medicinal herbs (various types), spices, specific dry food items (onion, garlic, potatoes and vegetables).

A second new license was issued for treatment of fresh onion / garlic / potatoes / sweet potatoes.

Currently, NCRRT is constructed an industrial food irradiation plant together with a medical product sterilization plant at the port of Alexandria. The buildings of the plants are under construction.

The following are some of the projects on food irradiation that are currently being carried out at NCRRT;

- Inhibition of sprouting of potatoes, onions and garlic
- Shelf-life extension of vegetables and fruits such as strawberries, tomatoes, eggplants, plum, apricots, figs, olives, etc…
- Radiation preservation of meat and meat products, and chicken and chicken products for increasing shelf life.
- Radiation preservation of fish, seafood and fish products.
- Wholesomeness of irradiated food.
- Elimination of pathogenic microorganisms from animal feed.
- Desinfection of store grains and their products.
- Identification of irradiated foods – in cooperation with institution in Berlin, Germany,
- Packaging system for irradiated food.
- Conversion of agriculture waste into animal feeds.
- Study of insects in relation to food irradiation.

Microbiology Department under the Biotechnology Division is also conducting some basic studies on microbiology in relation to food.

6.2. Radiation Sterilization of Healthcare Products is one of the main activities of NCRRT. Research activities on radiation sterilization of pharmaceuticals, medical and healthcare products are carried out by several Departments such as Radiation Biology Department, Health Research Department, Microbiology Department, Natural Product Department and supported other laboratories such as quality control laboratory for bio-burden, sterility test and dosimetry. Up to date there were more than the above Departments, which were supervised by more than 60 Professors and Associate Professors, produced 100 Ph.D. and M.Sc. thesis. The main facilities used for research work are gamma cells/chambers.

The commercial radiation sterilization activities started in 1980 as 500 boxes were sterilized using the Mega Gamma 1 type J-6600 irradiator with initial activity was 500 kCi. In 1997, the quantity of gamma sterilization products has increased to 19,000 boxes. The type and number of irradiated products also increased ranging from blood lines, droppers, kidney filters, Petri-dishes, aluminum foil, plaster dressing, dressing, valves, surgical gloves, cat gut-chromic, masks dressing, medical packages, catheters, medical preparations (antibiotics), syringes, needles, intravenous sets, pharmaceuticals products to medicinal herbs, spices and dry food items. Up to now, NCRRT has provided radiation sterilization services to more than 70 companies for more than 200 types of products. Currently an industrial sterilization gamma plant for medical products is being constructed at the port of Alexandria, Egypt.

6.3. Electron Beam Processing of Polymers, Biomaterials and Surface Curing of Materials is another field of research that has many applications. NCRRT is equipped with 1.5 MeV, 25 mA, electron accelerator that is in operation and in good condition. This electron accelerator is used mainly for radiation processing of polymers and biomaterials such as wound dressing hydrogels, stimuli-response membranes, and drug-delivery systems for control release as well as for surface curing of materials.

Polymer Chemistry Department and Radiation Chemistry Department are the two main users of the electron beam accelerators. Other departments that have interest on electron beam accelerator are Radiation Engineering Department, Solids and Accelerator Departments and Radiation Physics Department. Several testing laboratories such as Central Laboratory (mechanical, thermal and analytical equipments), Dosimetry Laboratory and Radiation Protection Laboratory support research activities on radiation processing of polymer, biomaterials and surface curing.

There are more than 30 Ph.D. and M.Sc. thesis generated from radiation processing research at NCRRT. At the Department of Polymer Chemistry and Radiation Chemistry, there are more than 50 scientific staff. The strength of the radiation processing research at NCRRT is on the following:

- Radiation grafting of membrane for various applications in industry, medical and agriculture such as for removal of detergents/pesticides, removal of toxic elements and separation of radio-nuclides.
- Radiation processing of hydrogels for wound dressing, drug delivery, magnetic/electric sensitive, hydrogels containing functional group for recovery metal ions, hydrogels for agricultural use – control release of pesticides.
- Recycling of waste polymer such as rubber and agricultural waste such as rice husk, cotton husk for making composites as shielding and container for transport.
- Polymer blend and composites such as rubber/polymer/glass fibers composite.
- Electron beam processing of polymeric industrial products such heat shrinkable products.
- Wood polymer composites and surface curing of coatings of wood products

Currently, some of the products generated from research activities are at the stage of commercialization. NCRRT is in negotiation with El-Sweedy Co. to irradiate heat shrinkable tubes at commercial scale. Meanwhile, NCRRT has entered into agreement with the pharmaceutical company for commercialization of the hydrogels for wound dressing. About 2,500 pieces of hydrogels has been distributed to several hospitals for clinical tests. In addition, 100 pieces of electron beam curing of laminated and surface coatings of wood has been distributed to several companies as promotional and demonstration of the electron beam curing technology.

In conclusion, electron beam processing of polymers, biomaterials and curing of surface coatings are very promising for industrial applications. NCRRT has started to produce products and to convince the industry to take up the research findings for commercialization. This achievement can be shared among the country in the region as well as neighboring countries.

7. Applied Industrial Activities at NCRRT

The major applied industrial activities at NCRRT are dealing with four main tracks, in the field of radiation processing of polymeric materials for industrial, agricultural and biomedical applications. The developments of polymeric materials are carried out by radiation technique includes radiation grafting, radiation curing, cross linking and degradation.

- Radiation graft copolymerization
- Blend and composite materials
- Radiation curing and surface coating
- Biomedical applications
- Radiation Effects on polymers (Cross linking and degradation)

7.1. Radiation Graft Copolymerization

Radiation-induced graft copolymerization of different monomers onto various polymers, using gamma and electron beam irradiations. The grafted and modified films may find varied applications such as ion-exchange membranes, reverse osmosis desalination of saline water, wastewater treatments from heavy and toxic metals, kidney dialysis, selective metal separation, protein adsorption and enzyme immobilization and heat resistant films.

7.2. Blend and Composite Materials

Blends and composite materials are prepared by radiation copolymerization of different polymers onto different industrial natural solid wastes. The recycling of some polymeric wastes are investigated and modified for the use as shielding for gamma rays and radiation sources. Modified and advanced copolymers containing various metal complexes are being prepared, which exhibit characteristic electrical and thermal properties for the purpose of its practical use as semiconductors and also for the removal of heavy metals and some organic or inorganic compounds as pollutants in the environment.

7.3. Radiation Curing and Surface Coating

The uses of electron beam irradiation in wood surface coating with unsaturated polyester for the protection and decoration purposes as marble like surface and crystal foam like glass in shape. Also, glass surface coating with synthetic adhesives for pigment fixation as drawing and printing. It is also used for the preparation of adhesive surface coating as pressure sensitive sheets for medical and industrial applications (self adhesive sheets). The rubber-shock pads from recycled waste rubber and thermoplastic polymeric binder have been prepared using electron beam at 50 kGy for sporting and athletic applications.

7.4. Biomedical Applications

The hydrogels are water-swollen network of hydrophilic copolymers or homopolymer. The use of hydrogels in biomedical has led to advances in biocompatible materials for artificial organ, methods of providing temporary support for wound, artificial skin as well as a base for drug carrier. Using hydrogel systems, of polymer composition and structure can have profound effect on diffusive properties and can be utilized in important biomedical, agricultural and pharmacological applications. Novel biodegradable polymer-drug delivery systems using gamma and electron beam irradiation have been investigated to modulate controlled drug release. Also, responsive polymer gels are prepared and have characteristic properties whose change in response to specific environment stimuli including temperature, pH, electric field, solvent quality, light intensity, pressure, ionic strength and ion identity or specific chemical triggers like glucose. Studies have been also conducted to provide polymeric materials with ant-microbial activity for the use as drugs or packaging materials.

8. Recent Research Achievements and Applications

8.1. Wound Dressing Hydrogel

Wound dressings are cross-linked polymer gels in sheet form. Using the electron beam irradiation at NCRRT with the collaboration of CID pharmaceutical Co. will produce such wound dressing. This wound dressing hydrogels are excellent for helping to create or maintain a moist environment. By increasing moisture content, hydrogels have the ability to help cleans and debride necrotic tissue. Children specially recommend dressings for skin injuries. This hydrogel don't adhere to the wounds, painless clean contaminated wounds and heal them without appearing crusts. Effective in hydrating wound surfaces and liquefying necrotic tissue on the wound surface. It is non-adherent and can be removed without trauma to the wound bed with minimal or no exudes, Soothing" effect, promotes patient acceptance, maintain a moist wound environment tissue.

8.2. Hydrogel in Agricultural Purposes

Ionic super absorbent Polyacrylamide copolymer prepared by Electron Beam Irradiation could be used as retaining materials in the form of: sandy soil conditioners, seed additives, seed coatings, and plant growth regulator/ or protecting agents carriers for controlled release. At NCRRT, Polyacrylamide hydrogel derivatives have been prepared with the capability of absorbing 400g to 1800g of water per dry gram of hydrogel. Acrylamide-alkali metal acrylate crosslinked polymers represent a recent advance in polymer technology for crop protection. The possibility for such materials to be used in agricultural fields as soil conditioner was evaluated. The prepared hydrogel aid in plant germination seedling establishment, and growth was stuided. The effect of polymeric soil conditioner on *Zea mays L. and bean* plant growth and crop yield has also been extensively studied. From economic point of view, such agricultural hydrogels was evaluated through the growth and other plant responses.

The hydrogel ability to enhance and control plant growth when some growth regulators or pesticides are incorporated into the hydrogel matrix was investigated.

8.3. Stimuli-responsive Hydrogels

Considerable attention has been drawn to hydrogels that usually undergo a volume phase transition in response to the external stimuli such as pH, temperature, electric and magnetic for industrial and biomedical applications such as drug delivery systems. Hydrophilic polymer containing functional groups suitable for external response such as ionic groups for pH and electric field were prepared. Promising results were obtained for the ability of such materials to be used as stimuli-responsive delivery system in some practical applications.

9. Currently Applied Research Activities

Wastewater treatment from heavy and toxic metals and industrial waste dyes and others, using adsorbents materials that are prepared by radiation grafting and copolymerization of different shapes; films, hydrogels, fibers, pellets, agriculture wastes (Straw Rice and Bagasse), etc.

- Preparation of stimuli-responsive hydrogels for biomedical applications (drug delivery under control release to specific sites).
- Water sorbent hydrogels as soil conditioners for agriculture purposes (active area and very promising practical results are achieved).
- Reverse osmosis membranes for desalination of seawater.
- Surface curing and coating of laminated woods and other materials using EB irradiation.
- Recycling of plastics and rubber wastes.
- Preparation of rubber-cement composites.
- Recycling of agricultural and industrial wastes.

10. International Cooperation with NCRRT

NCRRT has active international cooperation with the International Atomic Energy Agency (IAEA), Arab Atomic Energy Authority, IOM (Institute of Surface Modification), Leipzig, Germany; Julich-Germany; Takasaki Radiation Chemistry Research Establishment- JAERI, Japan; Polymer Institute, Budapest, Hungary; Hungarian Food Institutes; Canadian Atomic Energy Limited; Syrian Atomic Energy Commission, Saudi Arabian Universities etc. The cooperation consists of training/attachment of scientists and research projects.

Currently, NCRRT has two TC projects and three CRP with the IAEA in relation to radiation processing such as follows:

10.1. Technical Cooperation Projects (TC project)

- Upgrading the EB accelerator for industrial application(EGY/8/015)
- Establishment of High Dose Reference Laboratory (EGY/1/023)

10.2. Coordinated Research Projects (CRP)

- Radiation synthesis of stimuli responsive hydrogels and membranes for separation processes (11511/RO)
- The use of radiation processing sterilization or decontamination of pharmaceuticals and pharmaceutical raw materials (10353/RO).
- Use of irradiation to insure hygienic quality of fresh pre-cut fruits and vegetable and other minimally processed food of plant origin (302-D6-EGY-11680).

NCRRT has also received several IAEA trainees from Africa and Arab countries such as from Saudi Arabia, Syria, Jordan and Iran in the fields of polymer modification and improvements using radiation induced grafting and co-polymerization. In the past NCRRT has also actively involved in hosting many AFRA and IAEA training courses, meeting and workshop in relation to food irradiation, radiation processing of polymer and radiation sterilization.

With regards to commercialization, NCRRT annual income generated from the irradiation services in 2001-2002 is 0.7 million LE from medical products sterilization and 0.3 million from food irradiation. The irradiation services were provided to 57 medical companies and 20 food companies. This income is 60% higher than the previous year 2000-2001.

In 2003, negotiation with private companies is underway to commercialize hydrogels for wound dressing and to provide electron beam services for crosslinking of heat shrinkable tubes. Two new gamma irradiation plants will be constructed at the Alexandria City for medical product sterilization and food irradiation. The budget for installation of the cobalt sources for both plants has been approved and waiting for implementation.

Radiation Inactivation of Bioterrorism Agents
L.G. Gazsó and C.C. Ponta (Eds.)
IOS Press, 2005

Chemical, Biological, Radiological and Nuclear Terrorism: New Challenge for Protection and Crisis Management

Jiri MATOUSEK
Masaryk University Brno, Faculty of Science, EU Research Centre of Excellence for Environmental Chemistry and Ecotoxicology, Kamenice 3, CZ-625 00 Brno, Czech Republic

Abstract. Main three categories of chemical, biological, radiological and nuclear (CBRN) terrorism (often depicted as ultra- or superterrorism) according to form and material source are suggested. Differences between terrorism using weapons of mass destruction (WMD) and CBRN terrorism are explained, the WMD terrorism being only one of the three categories of CBRN terrorism.

CBRN terrorism involves in the first line misuse of the WMD, in the second line use of non weaponised toxic,contagious and radioactive materials, or primitive nuclear explosive devices.The third category implies violent strikes against infrastructures of present civilised and industrialised societies causing accidents with release of toxic agents, highly infectious materials and radionuclides resembling the pushing mechanism of disastrous wartime strikes with conventional weapons rather than peace-time accidents caused by personal, material or system failures.

Examples of already executed cases of CB-terrorism support this approach and categorisation of these highest forms of terrorism.

Introduction

Chemical, biological, radiological and nuclear (CBRN) terrorism (depicted also as ultra- or superterrorism) is often reported as terrorism using mass destruction weapons or WMD terrorism. This is however not quite correct reducing thus the wide spectrum of terrorist means and methods falling under this term (CBRN) only to military weaponry of this art. There are also other terminological misunderstandings considering certain violent events occuring in wars and armed conflicts to be terrorism. Terrorism as a violence or threat with violence of individuals or/and groups based on racial, national, ethnic, political, religious, economic, ecologic, sexual and other ideology or motivation against individuals or social groups predestines the choice of instruments of violence. Just the eve of the 21st century has marked escalation of violence in the shift from classical means (silent weapons, incendiaries, fire-arms, explosives) to actual inclusion of toxic and biological agents into the arsenals of terrorists. Increasing brutality of contemporary terrorism under its gradually proceeding internationalisation in globalised environment allows anticipating further development of terrorism from its classical forms using incendiaries, light weapons and explosives (that obviously never disappear) towards its most destructive forms – to chemical, biological, radiological and nuclear terrorism. By the way, even the terms *radiological* and *nuclear* are often deliberatery mixed even by some prestigious authors. Assessment of actual and potential forms and sources of CBRN terrorism is a necessary point of outcome for combating terrorism, effective preventive, repressive, protective, rescue and recovery measures and building respective systems on national and global scale.

This paper is aimed in the first line to categorise material sources and forms of CBRN terrorism. Differences between WMD and CBRN terrorism are elucidated. The former is considered to be no more and no less than one of the basic forms of the latter. Main forms of CBRN terrorism in wide possible varieties are reviewed and portrayed, generally far exceeding the alternatively used term WMD terrorism.

1. Material Sources (forms) of the CBRN Terrorism

Principal forms and their material sources shortly reviewed in this paragraph [1, 2], are portrayed in further paragraphs after profound analysis [3].

1.1. The First Source

The basic source of the CBRN terrorism (to which it is often terminologically reduced) is the misuse of military means, i.e. WMD. Possibilities of non-authorised using chemical, bacteriological (biological), toxin and nuclear weapons are limited mainly in the connection with their strategic importance and ongoing implementation of existing arms control and disarmament agreements.

1.2. The Second Source

This source (form) was applied at all known terrorist chemical and biological attacks and threats with attacks that have occured till now. This is own manufacturing WMD components, i.e. supertoxic lethal compounds, highly contagious bacteriological agents and toxins as well as misusing industrial toxic agents, infectious materials and radionuclides. High attention is devoted to the possible misuse of fissile materials, mainly of weapon-grade plutonium (WG Pu) and highly enriched uranium (HEU) for possible illicit manufacturing primitive nuclear device.

1.3. The Third Source

This form of CBR terrorism, till now generally underestimated, is the violent pushing of secondary effects typical for striking industrial and social infrastructures of modern society (nuclear, chemical, petrochemical and like) with conventional weapons in wars and armed conflicts. Such disastrous terrorist strikes causing sudden release of toxic, inflammable and liquefied chemicals, radionuclides and infectious materials (often with explosive character, sometimes followed by burnings, in some cases with fireballs) differ in the pushing mechanism from the similar but much less dramatically proceeding peacetime incidents and accidents, caused by personal, material or system failure or by natural forces.

2. Misuse of Military Means (WMD Terrorism)

CBRN terrorism based on using WMD means any misuse or unauthorised use of existing military arsenals of WMD, i.e. of concrete chemical, bacteriological (biological), toxin and nuclear weapons or their key components, acquired as a result of stealing, robbery or illicit trade from military bases, storage sites, manufacturing facilities, transports etc. (similarly like conventional weapons and explosives) which is however much more difficult because of items of strategic importance, strongly guarded. In this connection, the most probable is considered the access to already commissioned chemical, biological and nuclear weapons determined to elimination pursuant to respective bi- or multilateral arms-control and disarmament agreements in the States Parties or to standard weaponry in the non-States Parties (mainly in those in suspicions to support terrorist groups). In ongoing discussions, several reasons have been indicated why terrorists have not yet used WMD on mass scale, while during two decades (1979 – 1998) at least 12 conventional high-casualty assault cases are known involving more than 100 fatalities each, not to speak about the events of September 11, 2001 and after. Among the reasons, there are e.g. general reluctance to experiment with unfamiliar weapons and lack of corresponding precedents, fear that the weapon would harm the producer or user, fear that it would not work at all, or only too well, fear of alienating relevant constituencies and potential supporters on moral grounds, fear of unprecedented governmental crackdown and retaliation, lack of a perceived need for indiscriminate, high-capacity attacks for furthering goals of the group, and lack of funds to buy e.g. nuclear material on the black market [4]. Some of the reasons are however weakened due to increasing occurence of suicidal terrorist attacks. It is a question of time when the WMD terrorism actually emerges.

As for ***chemical weapons*** (CW) concerned, among the States Parties to the Convention on general and comprehensive prohibition of chemical weapons (CWC) (from 1993, that entered into force in 1997), main possessors are Russia and the USA and in the second line India and South Korea having declared small amounts of CW. Possession of CW is anticipated in some signatory states which have not yet ratified (e.g. Israel) and in others (mainly Arabic neighbours of Israel and their sympatisants and North Korea) that have even not yet signed, binding their signature on Israel's cancelling its nuclear programme.

In the case of ***bacteriological (biological) and toxin weapons*** (BTW), the situation is far less clear. On the one hand, there exists the Convention on prohibition, development, production and stockpiling BTW and on their destruction (BTWC) (from 1972, that entered into force 1975, by the way the first disarmament document outlawing one class of WMD) but on the other hand with the lack of any verification mechanisms. The States Parties only have declared elimination of the BTW stockpiles and transformation of corresponding R&D and production facilities to peaceful purposes. This Convention was signed in the time of ending classic era of BW (lasting for about six centuries) and starting rapid development of biotechnologies shortly after its signature. This is why some countries were continuing in the R&D of new BTW and defence against them. It is to be noted, that the use of BTW (similarly like the use of CW) in wars is prohibited by the Geneva Protocol (from 1925 that entered into force in 1928). As for toxin weapons (TW), their development, production, stockpiling and use are prohibited also by the above mentioned CWC, committing to their verified destruction. The long feeling absence of mechanisms for verification of eliminating BTW stockpiles will be solved by strengthening the régime of BTWC amending it by the Protocol on Implementation, elevating it on the similar level like CWC. The aim of this Protocol is to prove objectively and under international verification system that the stockpiles have been actually destroyed and any new ones are not being developed. This Protocol is a matter of very complicated negotiations proceeding from the early 1990s. It is a pity that the ongoing negotiations have entered a deadlock recently due to obstructions by the USA. From the point of view of potential targets of terrorism it can be added that while CW are only defined as antipersonnel and antianimal means, BTWC are defined also as means against plants. (Practice of use of toxic agents however involved also multiple intentional use of phytotoxic agents as a method of warfare e.g. in Malaysia in the 1950s and mainly in the Second Indochina War through the 1960s and early 1970s).

Radiological weapons (RW), i.e. intentional dissemination of radioactive materials in armed conflicts are not covered by any international agreement yet. During negotiations on this issue in Geneva Conference on Disarmament, this problem was withdrawn from the agenda in 1984 because the group of neutral and non aligned countries had required to binding this problem to the issue of nuclear weapons ban. Even from the military point of view such weapons were not considered actual (Their massive use gives neither quick effect like CW nor high delayed one like BTW). According to the general opinion, such weapons do not exist in military arsenals of WMD at all, even if the UNSCOM inspectors concluded in the early 1990s that Iraq had tried to develop such weapons.

Nuclear weapons (NW) as the most effective WMD are also subject of efforts aimed to their reduction and total ban as a final goal. Beside multilateral agreements involving testing and deployment in various environmental compartments and geographis zones, the most important is the Treaty on non proliferation of NW (NPT) (from 1968) closely connected with the system of IAEA safeguards. The core of nuclear arms control and partial disarmament are represented in the first line by bilateral agreements between USA and USSR (Russia respectively). Previous bilateral agreements concluded in the Cold War era on the strategic nuclear arms, such as nuclear bombers, ICBMs, SLBMs (including MIRVs) i.e. mainly SALT-I, SALT-II, did not reduce numbers of weapons, but only regulated bilaterally their balanced increase (limitation). This was the reason why the overall number of nuclear explosive devices gradually reached about 60 thousand (including nuclear charges of other NW states) worldwide in the mid-1980s. This was considered as actual nuclear multi-overkill and the nuclear status of the major Powers then created the situation characterised as Mutual Assured Destruction (MAD) which was not a strategy of any side as it has been sometimes reported but an objectively existing threat for both sides and for the whole mankind.The first document on actual nuclear disarmament is the bilateral agreement (USA – USSR) on the elimination of nuclear missiles with the range of 500 – 5500 km (IMF) signed in December 1987. In the early 1990s both sides commenced with actual reduction of NW what was reflected in the START-I and START-II agreements. Recent developments, mainly refusal of ratifying the latter agreement and unilateral US withdrawal from the ABM treaty of 1972 again have stopped ongoing reductions of NW arsenals. Moreover, recent orientation of this state towards mini- and even in micronukes with maintaining main elements of former nuclear strategy and build-up of the National Missile Defence (NMD) together with non-ratifying the Comprehensive Test-Ban Treaty (CTBT) gives bad signal also from the point of view of nuclear terrorism. The only promising act is the recent bilateral Moscow Agreement SORT (2002), containing programme of considerable gradual balanced reduction of nuclear arsenals of the USA and Russia. Contemporary estimated number of nuclear explosive devices is slightly over 20 thousand on the global scale. Simultaneously with growing number of NW states (USA, Russia, UK, France, China, India, Pakistan, Israel, and probably D.P.R.Korea) and rising number of states with missile technologies, the possibility of misusing NW increases. On the other hand, beside the mentioned legal constraints, there are also technological ones. Nuclear arsenals belong everywhere to a strategic interest of the highest importance, well protected against non-authorised use. Even the simplest aerial bombs are fitted with up to 5 independent security systems. Similarly, all nuclear means are fitted with independent circuits and locks (utilising the "more-persons" and multilevel security principles) to prevent them from unauthorised or accidental use or launching, the highest levels being controlled by the Command of Strategic Nuclear Forces, General Staffs and Presidents (USA, Russia) personally, in order to prevent accidental outburst of nuclear war.

3. Terrorism Using Chemical, Biological, Radiological and Nuclear Materials

Due to expected complications with stealing, robbery, illicit trade etc. of standard WMD for above mentioned reasons, some already executed terrorist attacks have shown the ability of potent terrorist groups to develop and manufacture supertoxic lethal chemicals and highly contagious biological agents [5]. This is enabled by the scientific and technological development and open access to information contributing also to the scientific and technological level of well organised terrorist groups. Growing dissemination of information, mainly by means of the global computer network enables quick global communication without risk of being disclosed. *Database of Incidents Involving Chemical, Biological, Radiological and Nuclear Materials, since 1900 till Present* in the Centre for Non-Proliferation Studies at the Monterey Institute of International Studies marked 329 cases till 1999. Most of them were connected with chemical and biological materials with clearly shown terrorist motivation.

Among ***chemicals***, beside manufacturing of toxic agents, in the first line the most effective supertoxic lethal chemicals (standard chemical warfare agents, like GB and VX), also misuse of stolen riot-control agents and of toxic industrial chemicals like chlorine, phosgene, hydrogen cyanide, cyanogen chloride etc. can be considered for direct mass-casualty or mass incapacitation attacks (to evoke panics). Much wider possibilities in the choice of chemicals are given in indirect strikes (or threat with strikes), i.e. through contaminated water or food. Beside stable supertoxic agents (like VX, HD etc.) many other chemicals like persistent pesticides, cyanides, compounds of arsenic, heavy metals, oil products etc. can be expected.

Biological agents of many types and origin, accessible e.g. from banks at medical and university institutes can be taken in consideration, including the misuse of infectious materials from foci of proceeding epidemies, both human and animal, not to speak about ability to manufacture some toxins.

Similarly, ***radionuclides*** from several peacetime sources including radioactive wastes, disseminated by various mechanisms, most probably through explosion could be used in terrorist attacks.

At the time being, diverse views have been expressed concerning the posibility of nuclear terrorism. Even if the construction of ***nuclear explosive device*** seems theoretically very simple, there are obviously large very qualified teams and very specific conditions necessary to develop and manufacture such device. Taking into consideration material and technological requirements including own safety, this is generally considered as hardly possible without state involvement. From strategic reasons, R&D and many manufacturing steps belong to a best guarded state secrecy, moreover localised in closed and strongly guarded areas. Nevertheless, taking in consideration proceeding nuclear proliferation, extent of the respective parts of the military-industrial complex in growing number of countries, partial escaping of information, material and brains are not excluded. Parallel to proceeding decrease of numbers of operational nuclear devices paradoxically increases possibility of terrorist misuse due to growing volume of fissile material from nuclear weapons decommissioned mainly according to above mentioned agreements (IMF, START-I) as well as due to routinely proceeding upgrading of nuclear arsenals. It is in the first line the weapon-grade plutonium (WG Pu) but more probably the highly enriched uranium (HEU) due to its extremely high amount and possibility to construct primitive types of nuclear explosive devices with the explosive yields typical for mininukes (yields in order of t – kt TNT eqivalent) resembling large conventional demolition charges. According to the latest analysis [6], the total amount of WG Pu in all NW states (i.e. inside NW plus considered and declared excess) is estimated to 251 tons, while similar figure of military owned HEU makes 1839 tons. It is assumed that WG Pu is much better guarded being the principal material for nuclear explosive items. While there has long been concern about nuclear material being acquired by non-state groups, recent reports indicate that NW may now, or soon will, be availabe to terrorist groups. Large quantities of HEU that are poorly controlled and otherwise unaccounted only in the former USSR and some other countries could be a very attractive source. This fissile material (only in the former USSR is enough to build about twenty thousand nuclear charges). HEU can, however, be readily diluted with natural uranium to a low-enriched level where it has high commercial value as proliferation-proof fuel for nuclear energetic reactors [7].

4. CBRN Terrorism through Violently Evoked Accidents

In various assessments of terrorist threats, this mechanism is more-or-less aside of analysing CBRN terrorism in spite of its high probable and even actual occurence and under certain circumstances also of very high effectiveness, sometimes less targetable, in other cases with large extent (which also belongs to the terrorist aims). The principle lies in violently evoked secondary effects of accidental acts, analogically as in cases of intentional and unintentional strikes with conventional weapons on infrastructures of modern civilised societies [8, 9, 10] such as chemical, petrochemical, nuclear, energetic, cooling and other facilities including social and hygienic installations. These terrorist strikes are aimed to release toxic, inflammable and liquefied chemicals, radionuclides, highly infectious materials (frequently accompanied with explosions, implosions, blast waves, fires with the effects of toxic products of burning, sometimes also with a fireball). These acts are similar to peacetime incidents and accidents caused by material, personal or system failures or to phenomena caused by natural forces (lightning, earthquake, volcanic eruption, earth or snow shift, deep frost, overflooding, high tide, tsunami and like) but differ very significantly in the extent and velocity of occurence of destructive factors due to pushing mechanism. So, e.g. a rupture in welding joint of the stationary or mobile tank filled with chlorine of other liquefied or highly volatile chemical or petrochemical or a leakage in large cooling equipment (food industry, ice-hockey arenas) filled with liquefied ammonia will be considerably different from hiting with e.g. anti-armour missile or explosive demolition charge. In the first case, typical for peacetime incidents, a longtime-acting point source with slow generating toxic plume is being formed. This enables relatively effective protective and rescue measures. The latter disastrous event, occuring in armed conflicts and in potential terrorist attacks, is represented by sudden dramatic creation of a momentum volume source with very quick evolution and proliferation of a plume possessing extremely toxic to lethal effects (depending on toxic chemical) within the close neighbourhood.

It is obvious that this category of terrorist attacks is applicable in the first line for ***chemical*** terrorism as mentioned above. One can however imagine also ***biological*** attack evoked through strike e.g. on storage of infectious waste or simply on the communal waste water purification facilities aimed to contaminate water supply etc. Very dangerous seem in this respect strikes against nuclear installations (hydrochemical uranium mining, enrichment and reprocessing facilities, nuclear reactors, storage sites for spent nuclear fuel and institutional radioactive waste, waste water and waste sludge reservoirs etc.), representing an extremely deleterious form of ***radiological*** terrorism with extensive and long-lasting contamination. This would be especially in case of a destroyed nuclear reactor much worse than the contamination following a nuclear attack (in spite of much higher initial activity of the latter) due to presence of nuclei with long half-time of radioactive decay in the reactor´s inventory.

To the category of evoked accident belongs as a matter of fact even the shocking scenario from the September 11, 2001. Intended accident of three Boeings 757 with suicidal steering against the twin WTC towers in New York and (allegedly) the Pentagon in Washington had the character of terminal phase of guided missile trajectory with extremely destructive power due to combination of kinetic energy of heavy objects flying a speed of above 500 mph and of thermic energy of tens of tons of burning jet fuel from nearly full tanks of aircrafts on the planned route to the western shore after relatively short flight from the Logan airport in Boston.

5. New Forms of Terrorism, Crisis Management and Protection of Population

New forms of terrorism, i.e. CBRN terrorism and their development on the background of globalised world have brought the possibility of sudden surprise extensive strike against social groups and social infrastructures anywhere and anytime. This is actually a new challenge of protecting population. In all countries, a system for protection of population has long been in existence. It was originally constructed just before the WW-II with the main emphasis on protection against air strikes and use of CW and further developed during the WW-II and the Cold War to respond to possible extensive use of conventional warfare and use of WMD. This system as a constitutive part of the comprehensive system of state defence was generally depicted as *civil defence.* During the 1960s – 1970s, in the time of the first *détente,* the global conflict between two major military-political alliances became highly improbable and new threats of modern civilised societies emerged. They were connected with steadily increasing capacities of production, storage and transportation units for toxic, liquefied, inflammable, explosive, radioactive and other dangerous materials. This was soon marked by increasing frequency and consequences of accidents in chemical, petrochemical, nuclear industries and open sea transportation by giant oil tankers as well as by crashes of military and space technological items, previously generally believed to be totally safe. The end of the 20th century was also accompanied by frequent elemental disasters such as earthquakes, large forest burnings, overfloodings, tsunamis, soil and snow shifts, extreme weather conditions, pandemics etc. These challenges gave rise of the new system of protecting population stressing the new threats the modern societies pose even if the military threats have not disappear. The previous systems were gradually transformed during the 1970s - 1990s which was generally reflected also in their depiction as *civil protection.*

The eve of the new century has been marked by profound changes in the organisation for protecting population to respond better, more quickly and in a more co-ordinated manner on the current threats the modern civilised societies pose. The main change e.g. in the Czech Republic, evolved from the new legislative acts concerning emergency planning and crisis management resulted mainly in the Integrated Rescue System, encompassing in an effective way and co-ordinated manner all elements taking part at preventive, repressive, protective, rescue and recovery measures and activities connected with all possible incidents, accidents and disasters that might emerge in peacetime as well as in wartime.

The terrorist acts are a very extreme form of violent threats posed to contemporary industrialised societies, sometimes similar in type like some industrial accidents but far more dangerous in surprise, velocity, intensity and range of effects and extent of consequences as measured by life losses and material damage. Emergency planning and management of prevention before, then repression, protection, rescue and recovery following terrorist attacks is therefore much more complicated that at any other peaceful incidents, accidents and even disasters of technogenic and elemental nature. Possible and probable terrorist attacks that can strike anywhere and anytime are an actual new challenge posing extremely high requirements on the whole system, emergency planning and preparedness to respond in adequate manner on actual site in correct time to at least considerably diminish the consequences of terrorist strikes.

6. Conclusions

Three main categories of chemical, biological, radiological and nuclear (CBRN) terrorism (often depicted as ultra- or superterrorism) according to for and material source are suggested and differences between terrorism using weapons of mass destruction (WMD) and CBRN terrorismare explained, the WMD terrorism being only one of the three principal categories of CBRN terrorism significantly limited due to strategic importance of WMD and also by ongoing implementation of existing arms-control and disarmament agreements. CBRN terrorism however involves in the first line misuse of the WMD (even if less probable), in the second line use of non-weaponosed toxic, contagious and radioactive materials, or primitive nuclear explosive devices. The third category, not yet fully assessed, implies violent strikes against infrastructures of modern civilised and industrialised societies causing accidents with release of toxic agents, highly infectious materials and radionuclides, resembling (in their pushing mechanism) disastrous wartime strikes with conventional weapons, rather than peacetime accidents caused by personal, material or system failure. Examples of already executed cases of chemical and biological terrorism support this approach and categorisation of these highest forms of terrorism. Current and developed systems for protecting population including the Czech Integrated Rescue System have to calculate with the possibility of these highly violent forms of terrorism and reflect them in the preventive, repressive, protective, rescue and recovery measures and thus in education, training and preparedness of the system and its individual constitutive elements, emergency planning, crisis management, as well as in the financial and material support, not to speak about necessary information, education and outreach on all levels of state and public administration, economic subjects and all segments of the civil society including all population layers and groups reaching literary every citizen.

References

[1] Matousek, J.: *International Politics* (Mezinárodní politika - in Czech) **25**, No 10, 19-22 (2001).

[2] Matousek, J.: *Civil Protection* (Civilná ochrana - in Slovak) **3**, No 4, 19-21 (2001).

[3] Streda, L., Matousek, J.: *Czech Military Review* (Vojenské rozhledy - in Czech) **43**, No 1, 98-113 (2002).

[4] UN Office for Drug Control and Crime Prevention, Vienna: 12 April 2002.

[5] Brackett, D. W.: *Holy Terror. Armageddon in Tokyo*, Weatherhill, New York, 1996.

[6] Pugwash Council: Statement on the dangers of nuclear terrorism, 11 November 2001.

[7] Schaper, A.: Terrorist use of nuclear weapons and control of weapons-usable materals. XV. International Amaldi Conference on Problems of International Security, Helsinki, 2003.

[8] Matousek, J.: *Scientific World* **33,** No 1, 24-29 (1989).

[9] Matousek, J.: *New Perspectives* **19**, No 4, 6-7 (1989).

[10] Matousek, J.: The release in war of dangerous forces from chemical facilities; in: Westing, A. (Ed.): *Environmental Hazards of War in an Industrialized World*, SAGE Publications, London - Newbury Park - New Delhi, 1990 pp 30-37.

Preventing is Better than Postfactum Intervention in Bioterrorism

Marian NEGUT
National Institute of Research-Development in Microbiology and Immunology "Cantacuzino", 103 Splaiul Independentei, 050096 Bucharest, Romania

Abstract. The terrorist anthrax epidemic reported in October – December 2001 in the USA demonstrated that biological weapon is possible to be used nowadays.

Difficulties in controlling potential biological agents (mainly dual use pathogens or genetically modified organisms) raise important concerns on the possible use in bioterrorism.

The preventive counter-bioterrorism strategy is based on:

- Promoting strong national legislation for preventing and combating bioterrorism;
- Adopting international regulations/recommendations against bioterrorism at national level
- Setting national plans and structures of biosecurity in human, animal and plant (bioprevention, biodefense)
- Controlling non-proliferation of biological weapon; controlling circulation of natural highly pathogenic organisms of natural or genetically modified agents as well as the size and types of different equipments for research and bioproduction activities.
- Integrated activities in public health: epidemic surveillance, suitable stocks of elected or new vaccines, network of well trained and equipped laboratories for a rapid detection and identification of microorganisms.

Introduction

The possibility of using diseases as a weapon was promoted for military purposes from the antiquity by the Scythes and later in the Middle Ages by the Tartars in Crimean war.

After discovering highly pathogenic bacteria an International Convention prohibiting the use of biological agents for military purposes was signed in 1923, but the 1925 Geneva Protocol prohibited the use of "bacteriological weapons" (as was the case with chemical weapons), but not the development, manufacturing and stockpiling of such weapons. So, in 1972 a new Biological Weapons Convention was promoted by the U.K., the USA and the former Soviet Union and more than 150 States Parties ratified this convention until 2003. A verification Protocol was initiated as a legal instrument for weaponisation control but working groups of specialists (microbiologists) and diplomats have not reached consensus on the definite form of this document. [8]

The terrorist attack against the USA in September 2001 and bioterrorist spreading of anthrax spores in October – December the same year (causing 22 cases and 6 deaths) alarmed not only the USA but also the international community. [2]

A new era of bioaggression was opened – the bioterrorism.

Table 1. Why the biological attack is preferred.

Biological weapon: arguments	Motivation	Remarks
- Easy to prepare	Spaces - reduced - difficult to be detected	- Rooms - Small Laboratories
- Low costs	Equipment - inexpensive - small volumes - dual use	- Easy to achieve - Less sophisticated than the known mass destruction weapons
	Reagents - inexpensive - common / large multipurpose utilization	- Available
	Aggressive agents - accessible from specialized laboratories even from authorized collection	- Some aggressive bacteria can be isolated from the environment or cases in man or animals
	Specialists - reduced number - short educated	No high professional education is needed
Storing	Accessible - small spaces - commercially available equipment Difficult to detect	No large amounts of organisms are necessary for terrorist attacks
Spreading	- A very large diversity - Very efficient in high density population and crowded places - They are long and highly resistant agents in the environment	- Contaminated "kami-kaze" in prodromal or clinical evolution can be used in highly contagious diseases
Huge impact - Medical/Public Health - Psycho - Social	- Very high morbidity and mortality (overcoming local/regional medical possibilities) - Uncontrolled spread as a result of the "evadation" of contaminated person - Panic is a common psycho-social reaction of the population producing social and economic disorder (disorganization)	- Panic is one of the most expected effect of the terrorist - Medical impact and social disorders generate and amplify economic difficulties
Security measures	- Protective measures: - very strict in the outbreak - on a very large area (depending on spreading) - very costly - Biological security: difficult to assure	-Important risk in handling and transportation

1. Why Bioterrorist Attack is Preferred?

Because bioterrorist attack is the most accessible and most attractive mass destruction weapon. This weapon is easy to prepare at low costs, is easy to use and has a huge psycho-social impact.

In Table 1 are summarized the main arguments for which biological weapon is preferred by terrorists.

2. Why Preventing is Better in Bioaggression?

Because a bioterrorist attack could be a disaster; the main reasons are presented in Table 2. [2, 4]

As medical reasons are the most important in bioaggression the most predictable direct target of a bioterrorist attack are high density human communities.

Table 2. Bioaggression could be a disaster: reasons.

Category of reason	Reasons	Explanations
MEDICAL	- Important and unpredictable morbidity / mortality - Overcoming of local or regional therapeutic capabilities - Huge costs for treatments and limiting the epidemic spreading	- Atypical clinical pictures → difficult diagnosis in early stages - Highly aggressive agents → high mortality - Viral, antibioresistant or genetically modified potential agents – no antibiotic treatment
SOCIAL	- Disorders in current economic and social activities - Panic = psycho-social reaction	- Important number of affected people missing from the activity - Uncontrolled psycho-social reaction happens frequently in disasters
ECONOMIC	- Important disorders in all sectors of economic activities (production, transport, commercial, services etc.) - Unexpectable effects in current community life - Unpredictable loss	By: - absenteeism - aberrant requirements - increasing demands in medical field
SECURITY	- Local, regional or even state security can be affected	- all mechanisms could be affected - disorders amplify the danger (all kinds)

Table 3. Some differences of military and bioterrorist aggression.

Bioaggression	Military	Terrorist
Significance	Limited – Nixon's Declaration 1969	High
Legal	Controlled – BTW CONVENTION – 1972 adopted by over 150 States	Legislation emerging
Probability	Very Low – even some states have declared or non-declared stocks	Very high accessibility and psycho-social impacts – important reason
Main implications	Military	Civilians

The implementation of most important measures of prevention in bioterrorism – as a part of bioaggression – seems to be difficult and costly, but who can really estimate the costs of a huge rapidly spreading epidemic of smallpox today for instance. [2]

For more than half a century the international community supported an enormous effort for limiting and eradicating the most aggressive transmissible diseases in man and animal. Any attack can be a disaster taking into account the lack or low level of immune protection of the population. Postfactum intervention could limit some effects but by far less than prevention of a bioterrorist attack. [2]

3. Preventing Bioterrorism

In many respects, military bioaggression differs from bioterrorism. In Table 3 are mentioned four of today differences as a result of changes in the balance of two kinds of bioaggression. [8]

Figure 1. PREVENTING BIOTERRORISM: NON-MEDICAL CONCEPTS

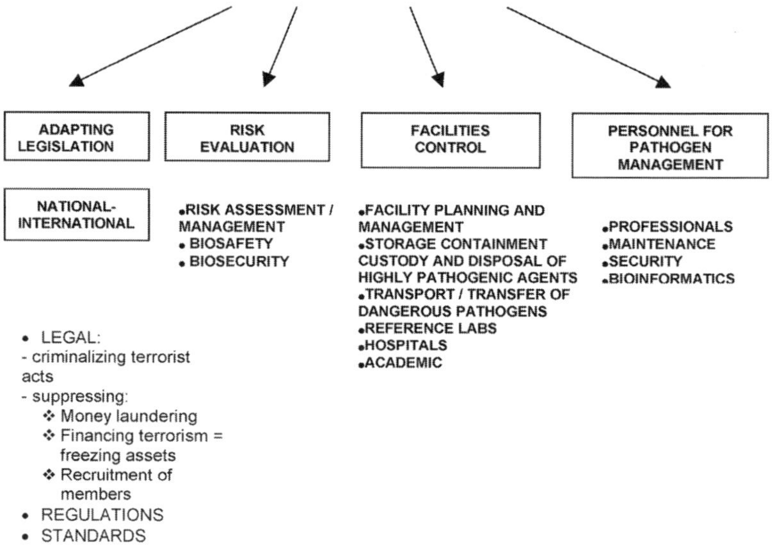

Two important differences have to be underlined. The prohibiting legislation in bioterrorism is basically national. Every country has to develop its own legislation to prohibit and to prevent local and international extending of bioterrorism. [5]

The main implication in bioaggression is civilian not military. Military help in preventing or combating bioterrorism is not excluded, but prevention and main implication belong to civilian authorities.

In preventing bioterrorism, two groups of concurrent non-medical and medical concepts can be established.

3.1. Non-Medical Concepts

In Fig. 1 are summarized the main non-medical concepts of measures regarding: legislation (to be nationally and internationally adopted) for prevention, risk evaluation, control of facilities and possibilities and personnel for pathogen management that can be estimated in preparing a bioterrorist attack. [3, 6]

Many countries promoted and introduced laws and regulations already for criminalizing and suppressing terrorist acts, bioterrorist especially.

Suppressing money laundering and financing terrorism are the most important measures internationally adopted for limiting bioaggressive proliferation. [3]

Many of the other measures of facilities control and risk evaluation are reinforced, adapted to new regulations corresponding to non-proliferation of dangerous pathogens.

Supplementary provisions have to be introduced as prohibitive measures not to be restrictive for technical and scientific development.

The particularities of dual use organisms and equipment as well as circulation of scientific information became a sensitive subject of debate between developing and technologically advanced countries.

3.2. Medical Concepts

The main medical concepts of preventing bioterrorism can be grouped in three interfering spheres of activities: bioprevention, bioprotection and biocontrol (Fig. 2).

Medical measures of bioprevention are presented in Fig. 3. Current epidemiological surveillance data referring to unusual epidemics (extension, evaluation), unusual clinical pictures or detection of an unusual etiologic agent are routinely reported. Medical and epidemiological records can be useful for early detecting and preventing the extension of a bioterrorist attack. In the same category of objectives, the monitoring of antibiotic use and resistance can suggest the presence of unusual organisms and can be an indicator of increasing stocks of necessary antibiotics. [1, 2, 4]

New promotion of vaccines as well as safety stocks are essential measures of bioprevention in bioterrorism.

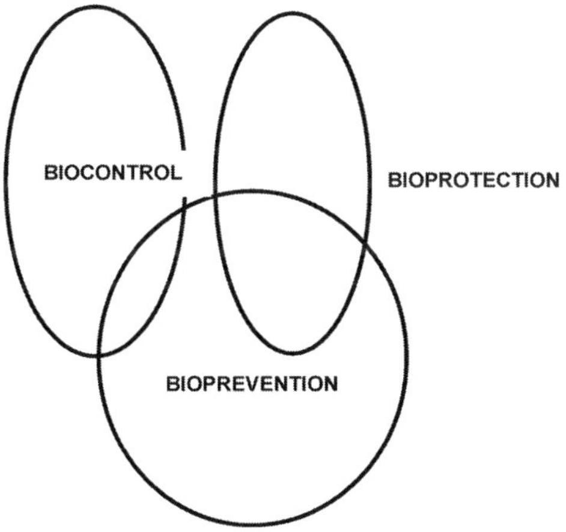

Figure 2. Preventing bioterrorism: medical concepts.

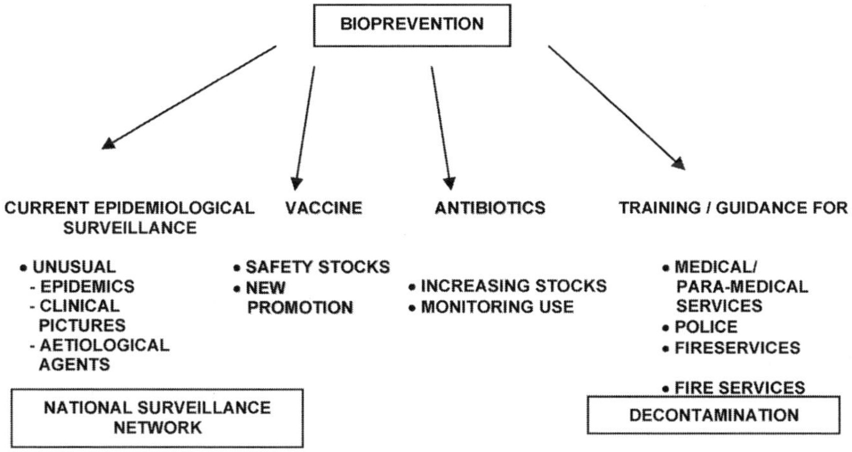

Figure 3. Preventing Bioterrorism: Medical Requirements.

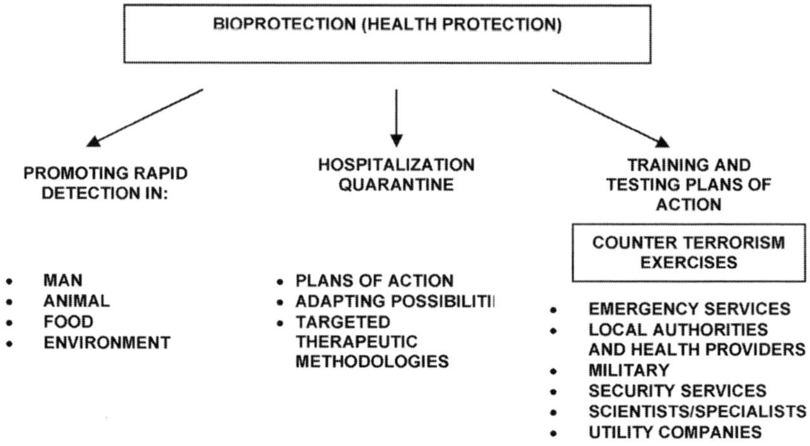

Figure 4. Preventing Bioterrorism: Medical Requirements.

Of course, no total protection is possible but limiting the epidemics and the dramatic evolutions of cases is a successful perspective.

Among some other reorganizational measures of medical and paramedical emergency services, special training of police and fire services is necessary.

Decontamination as a major antiepidemic measure is projected also as an essential measure in the biocontrol and bioprevention of bioaggression.

Bioprotection is the "task force" of emergency intervention. This operation can be successful only if there is a well established plan and preventing activities are well organized and checked periodically.

In Fig. 4 three groups of bioprotection (health protection) measures are mentioned: promoting rapid methods for pathogen detection, assuring facilities for hospitalization and quarantine and training plans of action.

Extended research for establishing new rapid methods of testing pathogens in man, food, animals and environment is necessary. Applying these methods currently in hospital and microbiology laboratories of national surveillance network is of utmost importance in bioprotection. [7]

Plans of action and counter-terrorism drills not only for medical specialists but also for the other public services, such as local authorities and health providers, security services, military and utility companies.

On hospital facilities during the bioterrorist attack depend qualified health, care duration, and costs of treatment and, the most important, survival rate. Any antiterrorist plan is based on hospital facilities that have to be suitably supplied with stocks of materials, mainly antibiotics, disinfectants and protection equipment for personnel and environment. [4, 7]

The last large sphere of medical requirements refers to biocontrol, more precisely to security measures.

The four important areas of biosecurity are mentioned in Fig. 5: biosafety manipulation and transfer of biologically dangerous products (natural or genetically modified pathogens), biosafety procedures for sampling and handling of pathological products or contaminated food and environment. Also surveillance of the release and disposal of biologicals is necessary as a security prevention measure.

Export control of well known pathogens listed as potential agents was introduced all over the world. National regulations are prohibitive but not enough efficient. [5]

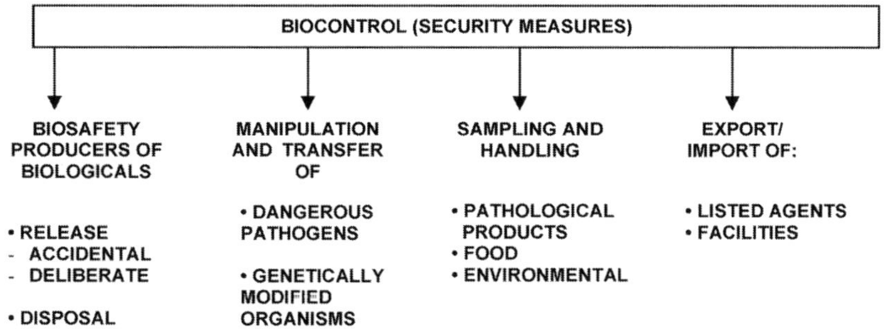

Figure 5. Preventing Bioterrorism: Medical Requirements.

The avoidance of customs barriers is possible as the necessary amounts of microorganisms to initiate and develop a bioterrorist action are small and easy to transfer.

Bioaggression can be internationally planned and initiated. But bioterrorism is nationally developed under unexpected forms, generating unpredictable human disasters. For this important reason preventing is better than postfactum intervention.

References

[1] Dando M.R., Klement C., Negut M. and Pearson S.G. (Eds.) – Maximizing the Security and Development Benefits from the Biological and Toxin Weapons Convention, *NATO Science Series, vol.38, Kluwer Academic Publishers*, Dordrecht 2002.

[2] Henderson A.D., Inglesby V.T., O'Toole T. – Bioterrorism, Guidelines for Medical and Public Health Management, *JAMA, USA,* 2002.

[3] Kellman B. and Muthe-Lindgren O. – National Laws and Measures for Counter-Terrorism Regulation of Biology, *The Program on Preventing Disease Weaponization, Strenghthening Law Enforcement and National Legislation,* August 2003.

[4] Paun L. – Infectious diseases, biological weapons, bioterrorism (In Romanian). *Ed. Amaltea*, 2003, Bucharest.

[5] Pearson S.G. and Dando M.R. (Eds.) – National Measures to Establish and Maintain the Security and Oversight of Pathogenic Microorganisms and Toxins, *Strengthening the Biological Weapons Convention, Briefing Paper No.4 (second series),* April 2003.

[6] Pearson S.G. and Dando M.R. (Eds.) – Maximizing the Benefits of the Inter Review Process: I: National Implementing Legislation, *Strengthening the Biological Weapons Convention, Briefing Paper No.6 (second series),* July 2003.

[7] WHO – Public health response to biological and chemical weapons, WHO guidance, Second edition, Prepublication issue.

[8] Wright Susan (Ed.) - Symposium "Current Problems of Biological Warfare and Disarmament", *Politics and the Life Sciences*, March 1999.

Radiation Inactivation of Bioterrorism Agents
L.G. Gazsó and C.C. Ponta (Eds.)
IOS Press, 2005

Potential Agents for Biological Weapons

Gábor FALUDI[a], Győző HORVÁTH[a,b] and György BERENCSI[c]
[a]*Institute of Health Protection, Hungarian Defense Forces*
H-1555 Budapest, POB. 68, Hungary
[b]*"Frederic Joliot-Curie" National Research Institute for Radiobiology and Radiohygiene H-1221 Budapest, Anna u. 5., Hungary*
[c]*"Béla Johan" National Center of Epidemiology*
H-1096 Budapest, Nagyvárad tér 2., Hungary

Abstract. Since October 2001, bio-agent risk has become a high priority threat factor and one of the key issues of the national and international anti-terrorist activities. The presentation is going to summarize known microbial agents in accordance with their potential or suitable for being used as biological weapons, and also the hazards or burdens attributed to the proliferation of such weapons or their bioterrorist use. Consisting mainly of natural pathogens, at this time various lists are known from different sources aiming to categorize microbial agents for this respect. Since 2001 WHO and CDC bio-agent classifications have become the most acknowledged and widely used ones. Such systemic collections of pathogens combined with the specific risk estimation viewpoints are not only theoretically useful for preventive military medical purposes, but it can directly and successfully assist NBC defence practices, too. Although in the era of genomics the real value of any lists is rather limited, yet they can provide useful references and practical guidance to the medical aspects of force protection planning, as well as for the entire spectrum of medical defence activities and efforts, including the development of newer and more efficient decontamination procedures.

Introduction

After a relatively short and silent period of the post Cold War era when only few extravagant scientists kept alive the alertness for the otherwise uninteresting and unbelievable issue of biological weapons [6] and - in general term - bio-agent hazard, recently the situation has dramatically changed due to the increase of asymmetric threats including the potential use of bio-agents by terrorists. Since the anthrax letter attack against US state officials and postal employees in October of 2001, the bio-agent risk has become a high priority threat factor and one of the key issues of the anti-terrorist international activities [15]. Now not only a wider array of experts, but politicians, decision makers, the public and certainly the media are also highly concerned about the consequences of such events. According to various computer assisted and scenario based threat assessments, nowadays all professionals agree that the malevolent or adversary use of bio-agents against military troops or highly dense civilian settlements is a real threat that could result in fatalities on a large scale in the targeted communities. On the other hand, even if well prepared and equipped, medical responders and the entire health care system would also be extremely burdened in the attacked area and neighbouring regions.

Based on the experience and knowledge acquired from natural outbreaks, our consequence analysis tools and forecast procedures for bio-attacks have significantly improved recently [26]. The main lesson we had to learn from the disastrous Southern African HIV/AIDS situation, the Ebola outbreaks, or the world-wide threat of SARS, and with respect of a possible outbreak of influenza A virus pandemic is that the epidemiological security of a country or a region must be regarded as an important and integrated part of the general national security.

In spite of the sporadic natural outbreaks, the general epidemiological security significantly improved world-wide in the last 40-50 years owing to the concerted efforts and actions of WHO and national health authorities in order to eradicate or control the most dangerous infectious diseases. As a result of the strict regulations, vaccination programs, and epidemiological surveillance, world-wide eradication of smallpox disease could be declared in the late seventies, also putting subsequently an end to the need for further vaccinations. Due to the overall beneficial tendencies, some of the most stringent countermeasures such as quarantine could also be omitted from the epidemiological repertoire. This comfort feeling was even reflected in the subsequent international recommendations as well as national legislation, resulting in the almost complete disappearance of the related knowledge from student textbooks or manuals.

This relatively long and peaceful Sleeping Beauty dream was interrupted by the recognition that biological weapons and bioterrorism - applying stealth technology - would present an enormous additional threat to natural occurrences by posing mankind to world-wide artificial outbreaks. Fortunately, the epidemiological situation regarding smallpox and the other most feared infections has not become worse generally, yet our attitude, risk perception and evaluation have dramatically changed during the recent years. Taking into account that the vast majority of human beings have become again susceptible to smallpox by now, we could easily find ourselves in a situation very similar to the one that occurred with the Amerindian nation in 1492. Furthermore, we must accept that the price of the epidemiological security has also significantly increased just like other things in the economy. If we intend to keep up with the positive side of life conditions, each nation has to build up a sort of complex and expensive preventive systems, to establish and accumulate stockpiles from antibiotics and vaccines that will hopefully never be used, and to develop more and more rapid methods for microbial diagnosis in order to control infectious diseases more effectively. At least for this respect, the difficult system of bio-defense is a definitive part of national and global security.

The aim of this work is to give a short review on the possible agents of biological weapons [2, 4, 23] and bioterrorism [8, 9], mainly from the aspects of the military preventive medicine. This discipline is strongly connected through many practical points to the biological weapons or biological defense preparedness, and for the army with special interest to force protection aspects. Therefore, the main focus points are as follows:

- Short introduction of BW terminology - as a „term of reference" - by running through some definitions.
- Comparison of several lists of known microbial agents that were already used or are believed to be suitable for biological weapon production.
- Summarizing those factors of natural or man-made origin that are thought to largely affect the potential of biological warfare agents.

1. "Terms of References"

Biological weapons (BW) contain highly pathogen microorganisms or their poisonous metabolic products (exo- and endotoxins) in order to cause diseases or kill humans, animals and plants, and also harm materials [10]. They represent a unique class of weapons of mass destruction (WMD), which are well distinguished from the other two ones - i.e. nuclear and

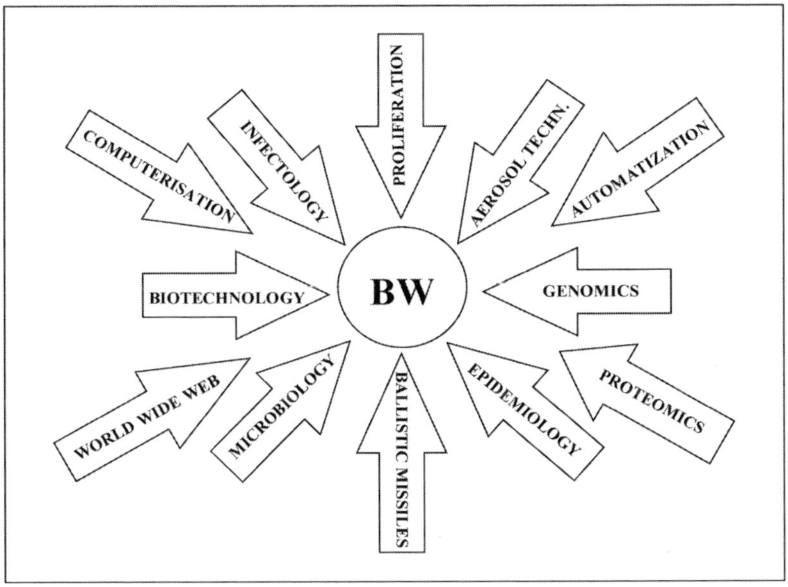

Figure 1. Influencing factors of biological weapon development.

chemical weapons - by the great diversity of the potential biological warfare agents and by their infectious nature leading to a large number of different and colourful clinical pictures. As far as the problematics of biological weapons is regarded, we are practically exposed to a problem that - at least partly – is related to the whole complexity of Mother Nature, but not of her best side. Additionally, the scientific-technical developments that have occurred recently in the related and background sciences such as biotechnology and infectology, industrial microbiology and genomics, together with the computerization and directing techniques of ballistic missiles via satellites, along with a wide array of civilization achievements – like the worldwide web - have resulted that this weapon family has grown up from its childhood by now (Figure 1.).

Biological weapons generally consist of three basic parts: (1) Delivery instrumentation representing the war-technical side; (2) Biological warfare agent, i.e. ready-to-use fighting material; and (3) Biological agents.

The war-technical side such as delivery instrumentation and methods of dispersion is beyond the objectives of this presentation.

Biological warfares refer to microorganisms or toxins in a properly processed and ready-to-use form suitable for effectively produce diseases or mortality in humans, animals, or plants. Implied in more or less sophisticated biological weapons or their covert and targeted use is able to call back the four equestrians of apocalypse from the Medieval Age to harm or kill large parts of civil populations as well as stuffs of military units. They can disrupt the attacked society, break down military fighting preparedness, overload the health care systems, and at last but not least, they can inhibit the sustainable development of the invaded territory for a long time. The influence spectrum of biological weapons has become pretty much expanded by their potential anti-material applications in order to cause biocorrosion or other damages in plastics of electrical circuits or fillings of military equipment. Biological warfare agents are composed of suitable biological agents as main constituents and different additives such as carriers, stabilizers, some wasting compounds, traces of applied antibiotics, and other remnants of the fermentation media.

Table 1. Consequence based classification of biological weapon agents.

	Incapacitation	Lethal
With epidemic potential	Influenza Venezuelan Equine Encephalitis (VEE)	Smallpox (Variola) Plague Haemolitic fever
Without epidemic potential	Staphylococcus Enterotoxin B (SEB) Trichotecene-2 (T2)	Botulinotoxin Anthrax

Biological warfare agents (BWA) are generally categorized by their consistence: simple ones are prepared from a single bio-agent like the plague or anthrax "bomb"-s, while complex ones are mixtures of different bio-agents. US cocktail(SEB+VEE+Q-fever) serves a historical example for this category. The mixed type was possibly produced by the former USSR, using biological as well as chemical warfare agents in the same load resulting the so-called and extensively discussed "yellow rain"(Sulphurmustard+sarin+t2). Biological warfare agents can be formulated for aerosol application in wet and dry forms.

Biological agents (BA) make the overall essence of biological weapons. According to the accumulated experiences achieved in microbiology, biological agents are cellular and non-cellular, natural or genetically modified microbial or other replicable entities or their toxic metabolic products. Classification trials aim to achieve a comprehensive, practical and useful typology of biological agents. However, the great diversity of biological agents and their various possible applications have resulted significant differences in the classification approaches which were reviewed by G. Pearson in 1997 [6]. He ranged bio-agents in larger blocks according to their characteristics such as pathogenicity, infectivity, lethality, replicability, etc.

Another approach to classification was the 2 times 2 squares method considering the expectable consequences (Table 1.). In this case, categorization is based on two important antinomy pairs in which the first pair is related to the expectable effects attributed to a biological threat and their final consequence. For this respect, agents can be lethal or cause incapacitation (i.e., severe degradation or entire lose of the physical or mental performance), with or without the potential of generating smaller or larger epidemic outbreaks. In this listing method weaponized agents are categorized by their common characters from the aspect of the final outcome such as secunder outbreaks everywhere, new zoonotic focuses in virgin niches, or massive persistent devastations. This approach can assist preventive staffs to estimate not only the extent of the direct effects, but to make shorter or longer epidemic prognosis and to recommend medical countermeasures. If we add a plus antinomy to the table like the curability of the given infection, it is also possible to convert the two dimensional table to a three-dimensional one. In this meaning smallpox is highly lethal and incurable, and also has a significant potential to generate epidemics.

This classification is in good accordance with Ken Alibek's recently published approach (Table 2.) [11]. Comparing the two lists, structuring similarities can be easily recognized. Nevertheless, the most appropriate classification as well as the best tool for the management of the diversity problem is provided by the scientific one, which is based upon the rigorous system of taxonomy. In this respect, prions present certain taxonomic problems in grouping, though their overall character meets the requirements for the biological agent definition. In other words, they are the causative agents of variant Creutzfeldt-Jacob's disease, i.e. they are natural, non-cellular, and replicative pathogens.

The microbial world has been sharing the globe with and adapted to mankind for many thousands of years. During their interactions, many of them developed an ability to become pathogen in the course of their parasitic coexistence with human hosts but only several spe-

Table 2. Partial Listing of known biological weapon agents by K. Alibek.

Lethal	Lethal/Incapaciting	Incapaciting
Anthrax	Brucellosis	VEE,EEE,WEEE
Bolivian haemorr.	Blastomycosis	Dengue fever
Ebola infection	Congo-Crim.hem.fev.	West Nile encephalitis
Glanders	Diphteria	Epoidemic typhus
Lassa infection	Psittacosis	Legionellosis
Marburg infection	SARS coronavirus infec	Murine typhus
Melioidosis	Japanese enceph.	Q fever
Plague	Omsk heamosshagic	Rift Valey fever
Smallpox	Russian S/S encephalitis	Influenza
Yellow fever	Tularemia	Monkeypox infection

Table 3. Requirements to be rated as "good" biological weapon agent.

- Lethal or incapacitating effect
- High morbidity and mortality ratios
- Epidemic potencial for transmission
- Low infectious dose
- High infectivity by aerosol dissemination
- Feasible for large scale production
- Storage, dissemination and environmental stability
- Target population susceptibility is general
- Difficult to identify
- Effective vaccination is unavailable or limited
- Available for production
- Ability to cause panic and social disruption

cies seemed to be suitable for satisfying the demands of weaponization. Military-industrial requirements for a "good" weaponizable bio-agent are summarized in Table 3 [1].

These requirements cover certain epidemiologic, military, and economical aims connected to the three main elements of epidemiology. A great number of conditions were stressed against agents, as stability in virulence and in the environment, in pathogenicity, lethality and so on. Important additive factor is the susceptibility of target population and the ability of invader troops to protect themselves.

2. The Lists

Based on the above requirements and assumed characteristics, time-to-time each army, large institutes and international organizations compile different top-secret, unclassified or open lists on the possible biological agents (Table 4.).

These lists can serve as starting points for risk estimation in preventive military medicine, and they can provide useful references and practical guidance to the medical aspects of force protection [14, 20, 21]. However, they have serious disadvantages, too. Even if revised and updated rather frequently, they only represent a time based cross-sectional situation, as every written text is static in comparison with the dynamism of life.

If we make comparisons between the known data of military biological weapon production in the past and the listed agents, one can easily recognize similarities as well as significant differences. A good example arises from the Iraqi BW project in which aflatoxins were

Table 4. Potential biological weapon agents (from NATO AMedP-6(B), Annex B, 1996).

Bacterial	Viral	Toxins
- Anthrax - Brucellosis - Cholera - Melioidosis - Plague - Q-fever - Tularemia	- Congo-Crimean Haemolitic Fever - Rift Valley fever - Smallpox - VEE	- Botulinum - C. perfringens toxin - Ricin - Saxitoxin - SEB

Table 5. Comparison of biological weapon and bio-terrorist agents.

Conventional biological weapon agents	Possible agents for bio-terrorism	
Pathogens		
Bacillus anthracis Brucella suis Coxiella burnetti Francisella tularensis Smallpox Virus encephalitis Virus haemorrhagic fever Yersinia pestis	Ascaris suum Bacillus anthracis Coxiella burnetti Giardia lamblia HIV Rickettsia prowaseki Salmonella typhi-murium Salmonella typhi	Shigella species Schistosoma species Vibrio cholerae Virus haemorrhagic fever Yellow fever virus Yersinia enterocolitica Yersinia pestis
Toxins		
Botulinum Ricin Staphylococcus enterotoxin B (SEB)	Botulinum Cholera endotoxin Diphteria toxin Nicotine	Ricin Snake venom Tetrodotoxin
Anti-crop agents		
	Rice blast Rye stem rust	Wheat stem rust

aimed to apply for this purposes. There are also available lists demonstrating the huge destroying capacity of other known biological weapon programs [17]. It is logically predictable to gain even more detailed knowledge about the proliferation of biological weapons when the possibly existing Asian projects will become available for studies hopefully in the near future.

The 21st century started with a sort of tragic events. The strength and potential of terrorism was clearly demonstrated to the world in 2001. People, who were still shocked by the consequences of the devastating terrorist attack on 11 of September, had to face a new type of danger right in October. Terrorism, as an advanced, planned and executed, and politically motivated violence took note of the destroying and extremely strong panic inducer capacity of WMD-s. The deliberate anthrax release contaminated large communal buildings, and anthrax polluted letters dramatically called the attention of the entire world to a new phenomenon: bio-terrorism [18].

Bio-terrorism has become a special and uncontrollable area of proliferation, as it evades all diplomatic treaties, legal and social rules by its inherent nature. By now, the problematic of the bio-terrorism related risk has become a global issue, as well as one of the primary focus points of the international anti-terrorist efforts. Defense aspects also differ from the regular military bio-defense because of the high number of the unpreventable victims, and

Table 6. Biological weapon agent list from the World Health Organization (Bacteria, Fungi and Protozoa).

Biological agents	**ICD code**	**Sources of data (see footnote)**						
		1	2	3	4	5	6	7
Bacteria (15 species)								
Bacillus anthracis	A22	+	+	+	+	+	+	+
Bartonella quintana	A79.0	+	-	-	+	-	-	-
Brucella species	A23	+	+	+	+	+	-	+
Burkholderia mallei	A24.0	+	+	+	+	-/+	-	+
Burkholderia pseudomallei	A24	+	+	+	+	+	-	+
Francisella tularensis	A21	+	+	+	+	+	+	+
Salmonella typhi	A01.0	+	+	-	+	+	-	-
Shigella subspecies	A03	+	-	-	-	+	-	-
Vibrio cholerae	A00	+	+	-	+	+	-	-
Yersinia pestis	A20	-	+	+	+	+	+	+
Coxiella burnetti	A78	+	+	+	+	+	-	+
Orientia tsutsugamusi	A75.3	+	-	-	-	+	-	-
Rickettsia prowaseki	A75	+	+	+	+	+	-	+
Rickettsia rickettsii	A77.0	+	+	-	+	+	-	+
Chlamydia psittaci	A70	-	-	-	-	+	-	-
Fungi (2 species)								
Coccidioides immitis	B38	+	+	-	-	+	+	-
Histoplasma capsulatum	B39.4	-	-	-	-	-	-	-
Protozoa (3 species)								
Naegleria fowleri	B26.2	-	-	-	-	-	-	+
Toxoplasma gondii	B58	-	+	-	-	-	-	-
Schistosoma species	B65	-	+	-	-	-	-	-

the different spectrum of the possible biological agents. Nevertheless, it was prognosticated much before 5th of October by public health experts and BW specialists that sooner or later bio-terrorism events were bound to happen [7]. After analyzing the databases of past events, spectrum differences between regular BW and bio-terror agents were re-evaluated and presented in 1999 [20, 22, 26, 17]. The challenge of bio-terrorism and the magnitude of the endangered civil population forced responsible public health and military specialists to re-assess biological agents and the real threat they present to the society.

Based on the former lists published by different institutions on BWs, two great centres, CDC (Centre of Disease Control) in the USA in 2000 [7], and the WHO (World Health Organization) in 2003 [3] summarized their actual viewpoints and compiled their own respective lists in their chem-bio guidance. As far as the WHO list is regarded, it required more than four years work of more than 100 experts from many international organizations and from different WHO regions to collect and compare former data sources and then prepare the current revised lists (Table 6.). Now it consists of more than 15 bacterial, 2 fungi, 3 protozoa, and 23 viral species.

CDC published a different classification system in 2000 (Table 7.). Taking into account all acceptable information, the established expert panel analyzed, selected and prioritized the possible agents of bio-terrorism - and the ones of chemical terrorism, as well. Based on their public health impact, dissemination potential and public perception, three blocks were formed from the agents and arranged in categories from A to C. The most dangerous agents – individually or publicly – were ranged as category A. Category B included less dangerous living agents and also toxins with the natural food and waterborne pathogens. Class C included the range of emerging infections.

Table 6. (cont.) Biological weapon agent list from the World Health Organization (Viruses).

Biological agents	ICD code	Sources of data (see footnote)						
		1	2	3	4	5	6	7
Viruses (23 species)								
Hanta viruses (KHF)	A98.5	-	+	-	+	+	-	-
Sin Nombre	J12.8	-	-	-	-	-	-	+
Congo-Crimean HF	A98.0	-	+	-	+	+	-	+
Rift Valley fever	A92.4	-	+	-	+	+	-	+
Ebola	A98.3	-	-	-	+	+	-	+
Marburg	A98.4	-	-	-	+	-	-	+
LCM	A87.2	-	+	-	+	-	-	-
Junin	A96.0	-	-	-	+	+	-	+
Machupo	A96.1	-	-	-	+	+	-	+
Lassa	A96.2	-	-	-	+	+	-	+
TBE	A84.0	+	+	-	+	+	-	+
Dengue	A90	+	+	-	+	+	-	-
Yellow fever	A95	+	+	-	+	+	-	+
Omsk HF	A98.1	-	-	-	-	+	-	-
Japanese encephalitis	A83.0	-	+	-	+	-	-	-
Western equine encephalitis	A83.1	-/+	+	-	+	-	-	+
Eastern equine encephalitis	A83.2	+	+	-	+	+	-	+
Chikungunya	A92.0	-	+	-	+	+	-	-
O'nyong-onyong	A92.1	+	+	-	-	-	-	-
Venezuelan equine encephalitis	A92.2	+	+	+	+	+	-	+
Smallpox	B03	-	+	-	+	+	+	+
Monkey pox	B04	+	-	-	+	-	-	+
Influenza	J10	-	+	-	-	+	-	-

Sources of Data: (1) UN Report to the Secretary General, New York, 1969. ; (2) World Health Organization, Geneva, 1970.; (3) UN Office of Dis. Aff., 1992.; (4) Australia Group doc., 1992.; (5) NATO AmedP-6(B) II., Brussels, 1996.; (6) APHA (the Chin) 17th. Edition, 2001.; (7) Ad Hoc Group/BTWC, 2001.

Table 7. Classification of biological agents by the CDC (USA).

Categories		
A	**B**	**C**
Bacillus anthracis Yersinia pestis Botulinum toxin Poxvirus hominis Francisella tularensis Haemorrhagic fevers (Ebola, Marburg) (Lassa, Junin)	Coxiella burnetti Brucella species Burkholderia mallei Alpha viruses/ VEM, WEE, EEE, Ricin C. perfringens/epsilon t. SEB Staphylococcus aureus E. coli O157:H7 Vibrio cholerae Cryptosporidium parvum	Nipah virus Yellow fever Hantaviruses TBE KHF Mr YBC

The advantage of this system lays in its advanced approach. It is a flexible and practical system containing also toxins and some other larger epidemiologic groups of bio-agents such as food-borne pathogens. Agents leading to enteric infections are easy to access, pro-

Table 8. Composite table of biological weapon agents (Bacteria, Fungi, Protozoa).

Biological agents	ICD code	WHO 1 - 7	CDC A – C	BSL 1 - 4
Bacteria (15 species)				
Bacillus anthracis	A22	1,2,3,4,5,6,7	A	3
Bartonella quintana	A79.0	1,4		2
Brucella species	A23	1,2,3,4,5,7	B	3
Burkholderia mallei	A24.0	1,2,3,4,5,7	B	3
Burkholderia pseudomallei	A24	1,2,3,4,5,7		3
Francisella tularensis	A21	1,2,3,4,5,6,7	A	2,3
Salmonella typhi	A01.0	1,2,4,5		3
Shigella subspecies	A03	1,5		2,3
Vibrio cholerae	A00	1,2,4,5	B	2
Yersinia pestis	A20	2,3,4,5,6,7	A	3
Coxiella burnetti	A78	1,2,3,4,5,7	B	3
Orientia tsutsugamusi	A75.3	1,5		3
Rickettsia prowaseki	A75	1,2,3,4,5,7		3
Rickettsia rickettsii	A77.0	1,2,4,5,7		3
Chlamydia psittaci	A70	5		3
Fungi (2 species)				
Coccidioides immitis	B38	1,2,5,6		3
Histoplasma capsulatum	B39.4	-		3
Protozoa (3 species)				
Naegleria fowleri	B26.2	7		3
Toxoplasma gondii	B58	2		2
Schistosoma species	B65	2		2

duce and apply for dissemination. On the other hand, effective water chlorination and rigorous hygienic restrictions can effectively decrease the predicted results of an intentional application. The CDC and WHO systems therefore are good tools for prevention and preparedness, as well.

From safety aspects, the majority of the viral agents were rendered in the higher 3 or 4 BSL categories, while bacterial agents have received a little bit lower 2-3 BSL ranging. Indeed, the WHO guidance gave us a fresh and sophisticated list of bio-agents for our practice. Table 8 represents a possible synthesis of these two comprehensive lists, and supplemented with the bio-safety ratings of agents.

3. Evolutionary and Man - Made Challenging Factors

Originally, the main challenging factor for "listing" biological agents was diversity, but by now two additional and more potent dangers should be taken into account [12, 21]. The first one is the group of those agents that are responsible for some newly discovered infections, in other words, the agents of emerging infections (Table 9). These agents – about 30 in 30 years – probably selected naturally as a result of the bubbling kettle of the evolution which has promoted changes in the living world for a long time and still affects every lifeforms in the Earth. In spite of the eventually launched such unfriendly products, it is clearly evidenced that this factory (kettle) still goes on working. Different strategies were selected under the pressure of evolution, as it is indicated by the examples of the highly changeable HIV virion or the peculiar behaviour of avian influenza strains in Asia and in Europe. This latter might be a good sample for the recognition and studying the machinery of the birth of a new Influenza A pandemic.

Table 8. (cont.) Composite table of biological weapon agents (Viruses).

Biological agents	ICD code	WHO 1 - 7	CDC A – C	BSL 1 -4
Viruses (23 species)				
Hanta viruses (KHF)	A98.5	2,4,5	C	3
Sin Nombre	J12.8	7		3
Congo-Crimean HF	A98.0	2,4,5,7	C	4
Rift Valley fever	A92.4	2,4,5,7		3
Ebola	A98.3	4,5,7	A	4
Marburg	A98.4	4,7	A	4
LCM	A87.2	2,4		2,3
Junin	A96.0	4,5,7	A	4
Machupo	A96.1	4,5,7		4
Lassa	A96.2	4,5,7	A	4
TBE	A84.0	1,2,4,5,7		3
Dengue	A90	1,2,4,5		3
Yellow fever	A95	1,2,4,5,7	C	3
Omsk HF	A98.1	5		3
Japanese encephalitis	A83.0	2,4		3
Western equine encephalitis	A83.1	1,2,4,7	B	3
Eastern equine encephalitis	A83.2	1,2,4,5,7	B	3
Chikungunya	A92.0	2,4,5		3
O'nyong-onyong	A92.1	1,2		2
Venezuelan equine encephalitis	A92.2	1,2,3,4,5,7	B	3
Smallpox	B03	1,4,5,6,7	A	4
Monkey pox	B04	1,4,7		3
Influenza	J10	2,5		2

Table 9. New products of evolution: Emerging infections.

1970s:	LEGIONAIRES DISEASE; EBOLA; LYME DISEASE; HANTAVIRUS; ROCKY MOUNTAIN SPOTTED FEVER
1980s:	TOXIC SHOCK SYNDROME; HIV/AIDS; LYME DISEASE; EHRLICHIOSIS; HEPATITIS C
1990s:	HANTAVIRUS PULMONARY SYNDROME; WEST NILE DISEASE; V. CREUTZFELDT-JACOB'S DISEASE
2003:	SARS (SEVERE ACUTE RESPIRATORY SYNDROME)
2004:	ALKHURMA, AVIAN INFLUENZA VARIANTS

The second source of the new possible bio-agents arises from the era of genomics. Many products of the former Soviet bio-weapon programs confirm the possible size of such danger. While the problems associated with natural biological agents have not been totally and correctly solved yet, now we have to reconsider the new threat derived from the genetically modified pathogens. The forth generation BW-s blew up the former limitations with the appearance of the genetically engineered agents [13, 19, 24]. The value of any conscripted BW agent lists also diminishes when we have to consider both in diagnostic and prognostic terms the presence of these genetically altered pathogens.

The altered microorganisms are artificially produced new species, strains or toxins. There are pretty much known routes for achieving such transformation: benign microbes can be supplemented to produce toxins or bio-regulators, or transform non-pathogens into pathogenic ones. Super-germs are known produced species for a stronger environmental resistance. It is also imaginable to modify the outer surface of agents so that they avoid detection or antibody neutralization. People seem to have passed a threshold at the end of the 20th century: the new molecular methods now offer a sort of unbelievable new agents that meet the requirements of weaponization. The range expanded from partial modification to brand new agents. Viral chimeras clearly demonstrate this trend [5].

Mankind is supposed to meet several extremely new agents even in the close future. Nevertheless, the great body of work that different committees and workshop participants put into the bio-agent listing will not be rendered entirely useless. It is a strong believes that lessons learnt in the school of natural pathogens can also be utilized in the management of the new "artificial" ones, as well as in new inventions like the express diagnosis of agents, the DNA vaccines, or the novel methods of decontamination such as ionizing radiation.

References

[1] R .Zajtchuk, R. F. Bellamy (ed.): Medical Aspects of Chemical and Biological Warfare. Textbook of Military Medicine part I. 1997. USA.
[2] E. Eitzen, J. Pavlin, T. Cieslak, G. Christopher, R. Culpepper: Medical management of biological casulties. 3. ed. 1998. Fort Detrick. USAMRIID.
[3] Public Health response to biological and chemical weapons. WHO Guidance. 2nd Ed. WHO. 2001. Prepublication Issue.
[4] Faludi, G., Rókusz, L.: A biológiai fegyver. 2003. MH. Egészségügyi Csoportfőnökség Kiadványa. (Biological weapon. Published by the Medical Directorate of Hungarian Defence Forces in 2003).
[5] K. Alibek, S. Handelman: Biohalál. Ármádia Budapest. 1999. (in Hungarian).
[6] M. Dando, G. S. Pearson, T. Tóth,: Verification of the Biological and Toxin Weapons Convention. NATO ASI Series. 2000. Kluwer Academic Publishers/Netherlands.
[7] CDC. Biological and Chemical Terrorism: Startegic Plan for Preparedness and Response MMWR 2000, 49. (RR-4).
[8] J. Miller, S. Engelberg, W.Broad: Baktériumháború. GABO. Budapest. 2002. (in Hungarian).
[9] Lukács A., Morvay P., Bioterror. - Új háború titkos fegyverekkel. Sprinter kiadó. (in Hungarian).
[10] Testimony of Prof. Dr. Keneth Alibek. 2002.
[11] K. Alibek: Smallpox: Weapon Threat and Disease. Smallpox BioSecurity Summit, Geneva. 2003.
[12] M. S. Bronze, M. M.Huycke, L. J. Machado, G. W. Voskuhl, R. A. Greenfield: Viral Agents as Biological Weapons and Agents of Bioterrorism. American Journal of Medical Sciences June, 2002. vol.: 323 No.:6 p.316-325.
[13] C. M. Fraser, M.R. Dando: Genomics and future biological weapons: the need for preventive action by the biomedical community. Nature Genetics 2001, vol.:29, 253-256.
[14] D. J. Kelly, A. L. Richards, J. Temenak, D. Strickman, G. A. Dash: The Past and Present Threat of Rickettsial Diseases to Military Medicine and International Public Health. Clinical Infectious Diseases2002, 34(Suppl. 4) S 145-169.
[15] W. F. Klietman, K. L. Ruoff: Bioterrorism:Implications for the Clinical Microbiologist. Clinical Microbiological Reviews 2001, Vol14, No.:2, p364-381.
[16] M. Leitenberg: The Biological Weapons Program of the former Soviet Union. Biologicals, 1993, 21(3)187-191.
[17] M. Leitenberg: Biological Weapons in the Twentieth Century: A Review and Analysis. Critical Reviews in Microbiology, 2001, 27(4)267-320.
[18] N. J. Beeching, D. A. B. Dance, A. R. O. Miller, R. C. Spencer: Biological Warfare and bioterrorism British Medical Journal 2002, Vol.:324, 336-339.
[19] M. Wheelis, M. Dando: New Technology and Future Developments in Biological Warfare. Disarmament Forum. 2000.4. 43-50.
[20] L. D. Rotz, A. S. Khan, S. R. Lillibridge, S. M. Ostroff, J. M. Hughes: Public Health Assesment of Potential Biological Terrorism Agents. Emerging Infectious Diseases 2002.8(2) 225-230.
[21] L. Borio et all: Heamorrhagic Fevers Viruses as Biological Weapons. JAMA 2002, 287(18) 2391-2405.
[22] M. G. Kortepeter, G. W. Parker: Potential Biological Weapons Threats. Emerging Infectious Diseases 1999. 5(4)523-527.
[23] Faludi Gábor: A biológiai fegyver jelentőségének megváltozása. Honvédorvos 1998.50(1) 37-69.
[24] C. M. Fraser, M. R. Dando: Genomics and future biological weapons: the need for preventive action by the biomedical community. Nature Genetics 2001, vol.:29, 253-256.
[25] W. F. Klietman, K. L. Ruoff: Bioterrorism:Implications for the Clinical Microbiologist. Clinical Microbiological Reviews 2001, Vol14, No.:2, p364-381.
[26] J. B. Tucker: Historical Trends Related to Bioterrorism: an Empirical Analysis. Emerging Infectious Diseases 1999. 5(4) 498-504.

Radiation Inactivation of Bioterrorism Agents
L.G. Gazsó and C.C. Ponta (Eds.)
IOS Press, 2005

Deployable (Molecular) Biological Laboratory: Concept & Reality

J. FŰRÉSZ[a], E. HALÁSZ[a], Á. NAGY[a], A. FŰRÉSZ[a], G. VESZELY[a], Zs. LAKATOS[a], J. FENT[a], Gy. HORVÁTH[a], T. VÉGHELYI[b], G. GYULAI[b] and L. NAGY[b]
[a]*Institute of Health Protection, HDF, H-1555. Budapest, POB 68. Hungary*
[b]*Technology Agency, MOD, H-1525. Budapest, POB 26. Hungary*

Abstract. Microbiological agents as well as bio-threats show some distinguishing and unique features if compared to other agents of mass destruction. For certain agents, lethality can be expected at concentrations far below the threshold of the conventional field detection capabilities. Their ability for multiplication instead of decomposition is very characteristic leading rather to the expansion than the attenuation of the consequences afterwards.

These specificities underline the importance of and need for a deployable bio-laboratory equipped with very rapid and sensitive methods. Until recently, when the development of recent molecular biological sciences gave an opportunity to introduce fast and very sensitive laboratory techniques, we have had only immunochemical techniques for these purposes.

To meet the requirements and as a first step, Hungarian Defence Forces developed and introduced a fieldable (molecular) biological-laboratory using up-to-date technology. The system is easily transportable and has all the parts needed for a complex molecular biological laboratory except housing. Readiness is reached 1.5 hours after deployment, and the first results are possible to get even in 3 to 6 hours depending on the number and characteristics (quality) of the samples. The whole staff consists of 4 persons. The mobile laboratory has been developed to detect hazardous agents from environmental and biological samples, and was successfully tested in several field exercises. In order to improve its bio-sampling and identification capabilities, as well as to become entirely independent from the hosting environment, the development of a mobile and fully containerized version has been initiated.

Introduction

The fast and safe confirmatory identification of biological warfare agents poses defence specialists to new challenge in many aspects. The key problem attributed to these agents derives mainly from their potential covered use. In case of the overt use of biological agents as weapons, the application itself calls the attention for danger and the concentration of the agent is usually high enough to carry out a successful analysis. The covert use however arises many problems, of which the need for continual screening, the low applied concentration and great diversity of agents and/or samples represent the biggest problem.

Calculating with 0.5-1 l breathing volume and with 12-15/min breath frequency, a soldier consumes 360-900 l air per hour. This means that even as low aerosol concentration as $3x10^4$ B. anthracis spores/m^3 can result in a lethal infection (the LD_{50} for human is between 8000-50000 inhaled spores). In case of more lethal agents (LD_{50} less than 10-100 microbes), even 3x10-100 microbes/ m^3 can lead to lethal infection.

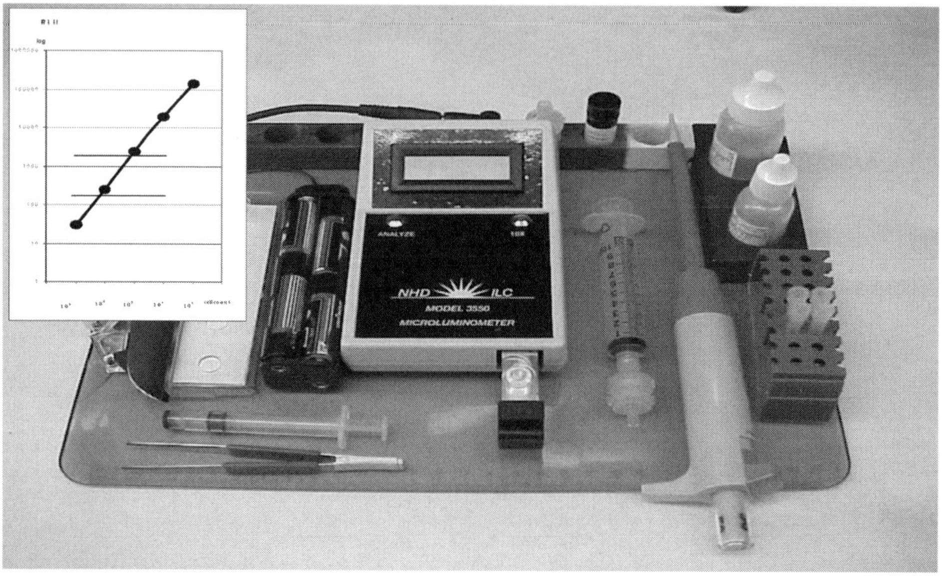

Figure 1. Hand held micro luminometer and its accessories used for ATP measurement. The inserted figure refers to our measurements where relative luminescence units (RLU) are plotted against the number of the colony forming units (CFU).

Some of the known measurement techniques are based on certain non-specific parameters of microbes such as the inherent fluorescence of living particles, or on the determination of compounds (chemicals) that are only produced by biochemical processes and therefore are suitable for making difference between living and non-living particles. A quite sensitive such technique is based on the luciferin-luciferase reaction in which a luminescence signal appears only when ATP is present. The signal correlates with the ATP concentration, which derives from the living cells (Figure 1) in the sample.

The application of chemical analysers for microbiological purposes appears to be one of the most promising directions. However, the sensitivity the most recent instrumentation still does not meet the expectations. The sensitivity of ordinary mass spectrometers is about 800 pg, and chemical ionisation mass spectrometers have a 20 pg sensitivity. On the other hand, viruses in a highly infective concentration of $10^4/m^3$ will represent approximately 1 pg which falls far below the maximum sensitivity of such equipment [1].

Another often used non-specific parameter is the size of the particles. For this purpose, laser light scattering procedures are commonly used. A special application of the laser light scattering technique is flow cytometry which, in addition to size measurements, let us also discriminate between microbes and common non-living particles in a buffer solution, either using special dyes like SybrGreen as DNA staining material, or even differentiate between living and non living organisms using special stains such as propidium iodide (Figure 2). The conduciveness of these non specific techniques is rather disputable due to the high background concentration of spores that can range between 10^5-10^7 spores/m^3

These drawbacks are partly compensated by the more specific techniques. The traditional culturing techniques for bacteria still remains the most specific golden standard. However, they are not applicable for toxins, and it is rather difficult to culture some microbial species, e.g. viruses. Microscopic techniques require a relatively pure suspension of the microorganism, and they cannot be used for toxins and viruses. The antigen and nucleic acid based techniques would resolve these problems, but some techniques does not meet the

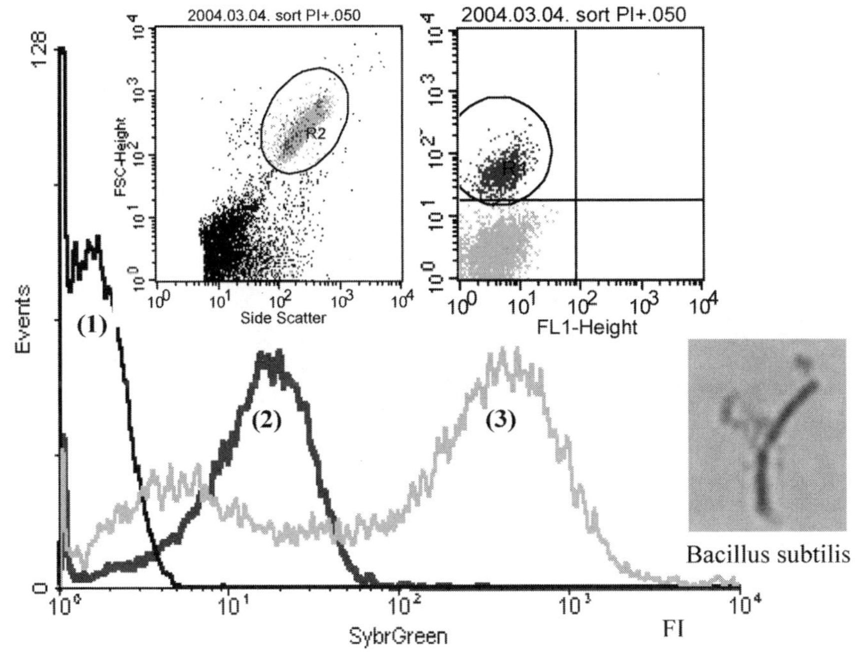

Figure 2. Flow cytometric determination of Bacillus subtilis DNA content by SybrGreen **(1)**: Buffer solution without SybrGreen; **(2)**: Buffer solution with SybrGreen; **(3)**: Bacterium suspension. Bottom right insert: photograph of B. subtilis. Top left insert: Forward and side scatterogram of B. subtilis; Top right insert: selection of living (grey) and dead (black) organism using propidium iodide for dead cell staining.

Table 1. Detection limits for selected identification technologies in case of Bacillus anthracis [8].

Method	Detection limit (CFU/ml)	Analysis Time
Handheld assay (HHA)	100,000	<15 min
Enzyme linked immunosorbent assay (ELISA)	10,000	90–180 min
Electrochemiluminescence (ECL)	1,000	90–180 min
Time resolved fluorescence (TRF)	1,000	90–180 min
Microscopic	10,000	5–120 min
Polimerase chain reaction (PCR)	1–100	<60 min
Culture	1–10	8–20 h

sensitivity requirements. The sensitivity of the regularly used techniques is summarized in Table 1.

The above mentioned detection limits seem to be very good. For the fast and reliable identification however, only PCR and immunochemical techniques [7] seem to be acceptable.

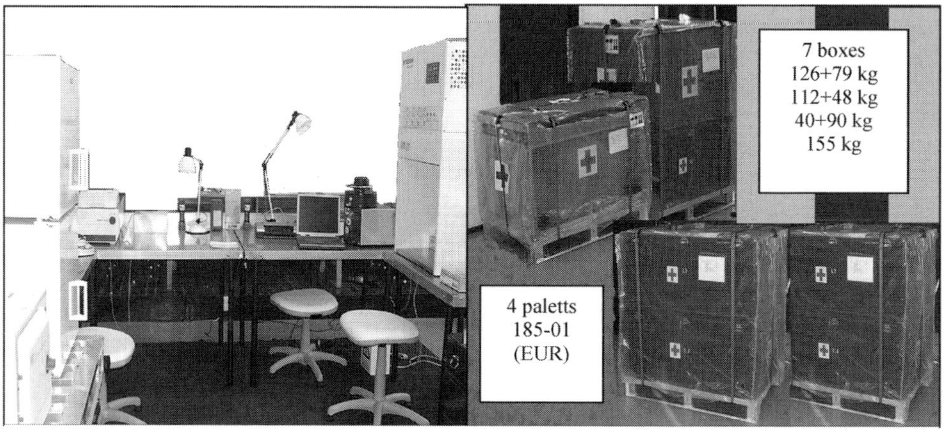

Figure 3. Deployable PCR laboratory in its unfolded and transportable (wrapped) condition.

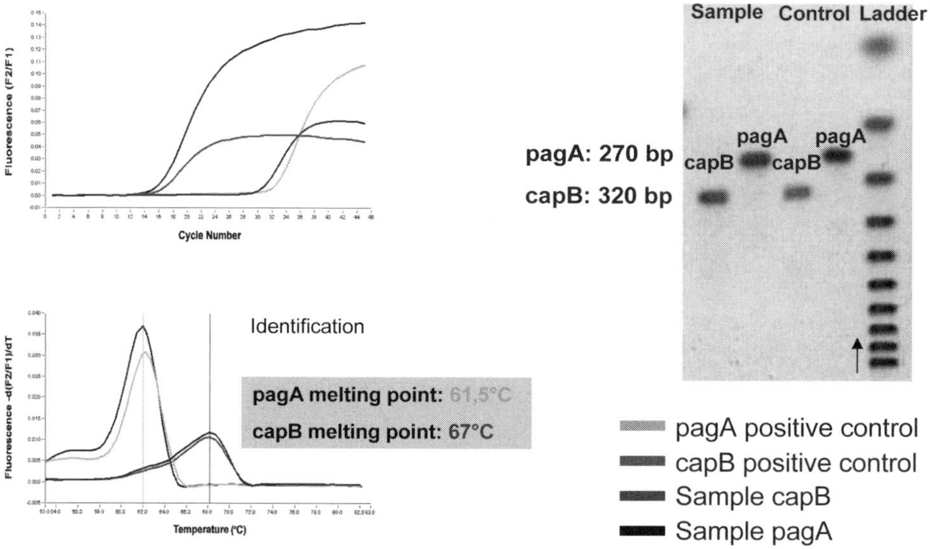

Figure 4. Detection and identification of Bacillus anthracis by real time PCR (Light cycler: Roche) Top left: The earlier the fluorescence starts to rise upward the greater was the concentration of the microbes in the sample. Left down: The melting point analysis proves the identity of the product. Right: Electrophoresis of the product proves the homogeneity of the sample and also proves the product identity.

1. Development of a Deployable Real Time PCR Laboratory

Recent developments in molecular biology also resulted in the appearance of highly sophisticated, fast and reliable techniques of which real time PCR is the most known and most acknowledged [2, 3, 4]. This technique lets the user to detect and characterise (identify) as few as 1-10 cells. However, the application of this method requires safe sample handling, effective sample preparation, and also clean environment for sample manipulation. For this purpose an enfolded laboratory was developed (Figure 3.).

Figure 5. Deployable rapid bio-analytical laboratory.

For the deployment, it was necessary to provide a closed, 25 m^2, tempered territory equipped with electricity. According to the requirements, deployment time is approximately 1 hour, the preparedness for the measurements takes 45 min, while sample preparation and performing the measurement need additional 1½ to 4 hours [5, 6]. The laboratory is equipped to handle environmental samples and is able to identify more than 25 microorganisms by 4 troops.

This laboratory has taken part and tested at numerous NATO field exercises. One of the representative results is summarised in Figure 4.

2. Development of a Deployable Laboratory for Confirmatory Identification

The need for an entirely independent laboratory resulted in the development of a housing equipped with electricity, air conditioning, regulated internal air pressure, internal and external safety systems. In working rooms, built in safety boxes provide safe working conditions for the personnel. The automatic DNA isolator, RT-PCR and electrochemiluminescence techniques support the capability for fast and sensitive confirmatory identification.

The component parts of the laboratory represent an overall modular system that is suitable for tailoring different laboratory settings according to the actual tasks and just before the deployment (Figure 5.).

The laboratory provides safe, level III conditions for sample handling, effective sample preparation and clean sample manipulation. For the deployment a quite flat territory is required. Deployment time is approximately one and half hour, the preparation for measurements takes additional 45 min. To reach entire readiness, approximately further 4 hours are needed. Time requirements for sample preparation and measurements are the same as mentioned before (1½ to 4 hours). The laboratory is prepared for handling environmental samples and is able to identify more than 25 microorganism species by 5 troops.

References

[1] Jim Gillespie (jgillesp@arl.army.mil).
[2] Detection of spores of Bacillus anthracis using the Polymerase Chain Reaction J.Infectious Diseases, 1992, 165, 1145-1148.
[3] M. Carl, R. Hawkins, N. Coulson, J. Lowe, L. Robertson, W. M. Nelson, R. W. Titball, J. N. Woody: Utilization of the rpoB gene as a specific chromosomal marker for real-time PCR detection of Bacillus anthracis.
[4] M. A. Lee, G. Brightwell, D. Leslie, H. Bird, A. Hamilton: Fluorescent detection techniques for real-time multiplex strand specific detection of Bacillus anthracis using rapid PCR J.Applied Microbiology, 1999, 87,218-223.
[5] M. Stemmler, H. Meyer: Rapid and specific detection of Coxiella burnetii P149-154.
[6] J. Zhou, M. N. Bruns, J. M. Tiedje: DNA recovery from Soils of Diverse Composition Applied and Environmental Microbiology,1996, 62 No 2, 316-322.
[7] Gatto-Menking D.L., Yu H., Bruno J.G., Goode M.T., Miller M, Zulich A.W.: Sensitive detection of biotoxoids and bacterial spores using an immunomagnetic-electrochemiluminescence sensor Biosens Bioelctron 1995 1010 (6-7):501-7.
[8] Leonard F. Peruski, Jr. and Anne Harwood Peruski: Rapid diagnostic assays in the genomic biology era: detection and identification of infectious disease and biological weapon agents BioTechniques, 2003, Vol.35 No 4, 840-46.

Irradiation Decontamination of Postal Mail and High-Risk Luggage

Marc DESROSIERS[a], Bert COURSEY[a], Stephen SELTZER[a], Lawrence HUDSON[a], James PUHL[a], Paul BERGSTROM[a], Fred BATEMAN[a], Sarenee COOPER[a], Douglas ALDERSON[a], Gregory KNUDSON[b], Thomas ELLIOTT[b], Michael SHOEMAKER[b], Joel LOWY[b], Stephen MILLER[b] and John DUNLOP[c]
[a]*Ionizing Radiation Division, Physics Laboratory, National Institute of Standards and Technology, Gaithersburg, Maryland, USA*
[b]*Armed Forces Radiobiology Research Institute, Bethesda, Maryland, USA*
[c]*United States Postal Service, Merrifield, Virginia, USA*

Abstract. In October 2001, first class letters, which were laced with *Bacillus anthracis* spores, were sent to political and media targets resulting in five deaths and 22 illnesses, significant mail service disruption, and economic loss. The White House Office of Science and Technology Policy established a technical task force on mail decontamination that included three key agencies: the National Institute of Standards and Technology (NIST); the Armed Forces Radiobiology Research Institute; and, the United States Postal Service. A cooperative effort between this task force and industry led to protocols for the processing of letter and parcel mail.

Currently, NIST is examining the technical issues and barriers to the use of ionizing radiation to mitigate bioterrorism agents in high-risk passenger luggage. The purpose of this work is to develop irradiation specifications, procedures, and protocols that will ensure that broad classes of bioterrorism agents in passenger luggage will be neutralized without damaging luggage contents and inconveniencing passengers with long delays. This work focuses on three areas: the assembly of critical input data, the development of a coupled computational-experimental verification approach for estimating the radiation dose that can be delivered to passenger luggage and the application of the computations to a larger variety of luggage configurations followed by the development of specifications, procedures, and protocols for the irradiation of passenger luggage.

An analysis of the expectations for growth in these and other homeland security areas where irradiation technology can be applied will be discussed.

Introduction

Shortly after the tragic events of September 11, 2001, the terror triggered by the anthrax-causing spore attacks affected every American. Since the postal service is integrated into daily life, the effects were profound. Within the Federal government, these attacks unleashed a flurry of activity aimed simultaneously at the remediation of the exposed facilities and the search for prevention/mitigation methods to halt further attacks. Government agencies were mobilized to address the current emergency as well as for guidance in formulating long-term policy changes. It was soon realized that ionizing radiation was well suited to eradicate the weaponized *Bacillus anthracis* spores dispersed in the nation's postal system. The Ionizing Radiation Division (IRD) of the National Institute of Standards and Technology (NIST) was uniquely positioned to bridge the gap between industry and government. As a Department of Commerce agency, NIST's mission to promote and develop measurements, standards and technology enables it to have a close working relationship with their industry stakeholders. In this role, the IRD offered to guide other Federal agencies to the irradiation technology best suited to its needs. The IRD's established trust and credibility with the private sector would also aid in the coordination of the people and facilities required to sanitize the mail in rapid fashion. Accordingly, NIST continues to play a key role in related studies such as the irradiation treatment of high-risk passenger luggage.

1. Mail Sanitation

1.1. Bacillus Anthracis Spore Attacks

NIST quickly ascended to a leadership role in a rapid succession of meetings organized by the President's Office of Science and Technology Policy (OSTP) and the United States Postal Service (USPS). Several NIST staff were assembled to brief a broad range of Federal agencies on industrial radiation-processing technology. NIST's intimate knowledge of the irradiation industry led to invaluable guidance in assessing a safe and cost-effective course of action. Since the IRD operates a national calibration service that routinely certifies industrial irradiation facilities, it was able to attest that they were fully capable of sanitizing the mail with the highest level of quality possible. However, for manufactured products that are commonly sterilized with ionizing radiation, the individual items are identical and their packing within an irradiation container is uniform. Because postal mail is heterogeneous in nature and there was no standard packing pattern or density for the mail, there was a need to demonstrate the efficacy of the process.

As the process had begun in late October 2001, the OSTP formed a task force to develop a plan to validate the irradiation protocol proposed by the Titan Corporation[1] irradiation facility personnel. The technical task force included three key agencies: the National Institute of Standards and Technology; the Armed Forces Radiobiology Research Institute (AFRRI); and, the United States Postal Service. NIST suggested that a series of test-mail letter boxes be prepared. As the technical leaders of the task force, NIST dosimetry experts partnered with spore biologists from AFRRI to design the test. Within 24 h a procedure was agreed upon, and preparations to assemble the test boxes had begun. It was decided to prepare three letter-mail trays in a manner identical to the contaminated mail. This would require that letter mail be packed in "MM" type trays then wrapped/sealed in a biohazard plastic bag. This bagged tray would be inserted into the MM-tray cardboard sleeve and wrapped/sealed a second time in a biohazard plastic bag. The first test-tray box would contain dummy mail with NIST and AFRRI indicators that would be hand carried to the irradiation facility by task-force representatives. A second box was assembled in an identical manner with the exception that this box was marked with an internal label identifying it as containing experimental artifacts; it was shipped through the Brentwood facility with the contaminated mail from there. A third box was prepared and shipped in the same manner as the second; however, this box included a number of non-paper objects in separate articles of dummy mail: coins, CDROM, floppy disk, plastic sheets, metal sheets, and metal paper clips. After the first box was flown to the irradiation facility and processed, the dosimeter materials were removed and measured at NIST while the spore indicators were measured at AFRRI. After the analysis period, and exactly one week from the project's conception, NIST and AFRRI reported to the OSTP and USPS that the tests confirmed that the process

---

[1] The mention of commercial products throughout this paper does not imply recommendation or endorsement by the National Institute of Standards and Technology, nor does it imply that products identified are necessarily the best available for this purpose.

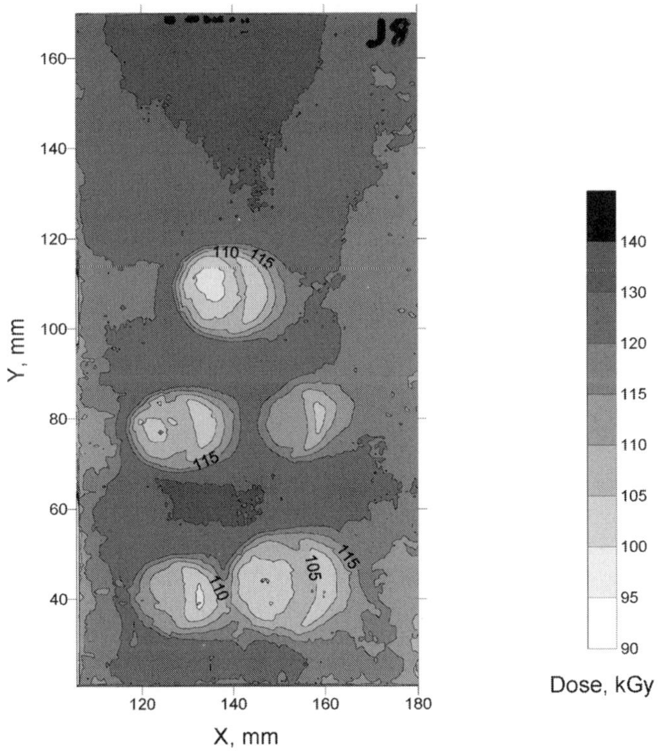

Figure 1. Radiochromic-film dose map mounted against coins in an envelope. The dark marks at the top of the image are for identification.

was safe. In turn, the OSTP Director endorsed the process and recommended that NIST continue to actively monitor the activity.

The irradiation process is certified by alanine-EPR dosimetry. Alanine dosimeters are of the highest metrological quality and are used by national metrology institutes worldwide to operate their dosimetry services. The uncertainty at the 95% confidence level for alanine pellets is ≈ 2% and for alanine films ≈ 3%. The absorbed doses measured in the Titan-irradiated test boxes were typically double the target minimum dose; some dosimeters measured triple the target minimum dose. Also, for some dosimeters it was obvious that they experienced temperature at or above their melting point (≈ 85 °C). Efforts to modify the protocol to control overdosing are discussed later in this section. Objects (mentioned above) placed into one of the test boxes yielded interesting results. For these tests a radiochromic dosimetry film was used for the measurements; the accuracy and precision of this system is lower (≈ 5% at 95% confidence) than the alanine system, but the radiochromic film offers two-dimensional mapping of the energy deposition. The films were hermetically sealed in a foil pouch (ambient light and relative humidity affect the results) and placed against the objects inside paper envelopes. The objects did not significantly shield the material behind it (Figure 1). This was due to the fact that the boxes are flipped 180° midway through the treatment. The floppy disk was not readable because of damage to the plastic case; however, although the CDROM jewel case was severely damaged, the CDROM was readable. Figure 2 shows the dose distribution within the plane of an envelope irradiated end-on (electron beam parallel to the long side of the envelope plane).

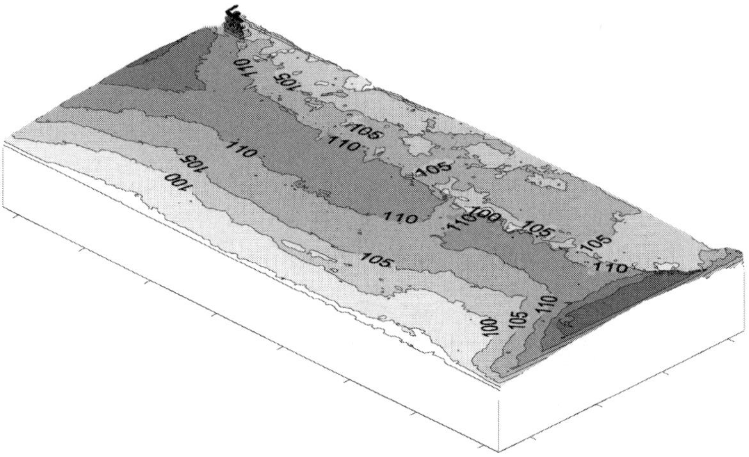

Figure 2. Radiochromic-film map of dose distribution within the plane of an envelope (doses in kGy). The dark mark in the upper corner of the image is for identification.

Shortly after the mail sanitation work began, it was realized that the throughput of the Titan facility coupled with the large volumes of contaminated mail at the Brentwood and Trenton mail facilities would cause unacceptable delays in the mail-delivery times. Moreover, in addition to letter mail, so-called "flats" or large envelopes and magazines, along with packages of larger dimensions, had to be processed. The electron-beam accelerator characteristics for this particular Titan facility would not be appropriate for the processing of large packages. Thus, a second facility, Ion Beam Applications (IBA), was contracted to irradiate mail. For the Titan facility letter mail was stacked vertically into trays and the trays were stood on end and moved through the horizontal electron beam. The IBA facility used a vertical electron beam, and the mail was stacked flat in the tray to avoid any issues with the shifting of contents when the mail trays were flipped for a required second pass through the beam. Because the IBA facility design was different than that for Titan facility, this process was once again validated. The alanine dosimeters in these test boxes measured doses predominately 50 % to 100 % greater than the target minimum dose. These data are much more typical of the dose range experienced in industrial processing. The absorbed dose measured in the center of a stack of flats was about 50 % greater than the target minimum dose.

Operating in an emergency mode from the onset, mail irradiation did not enjoy the normal period of planning, design and testing that would optimize the process from the perspectives of the irradiation process and product quality. To facilitate this, NIST acted as an intermediary between the processors and the packers. To achieve product consistency, NIST worked with the USPS to gather feedback on the product quality to formulate packing guidelines, and then coordinated this with industrial irradiation facilities by setting acceptance criteria for letter trays. Some of the early inconsistencies in packing coupled with conservative irradiation settings led to an over-irradiated product. The chemical degradation from the combined effects of radiation, heat, steam and ozone, produced undesirable physical effects in the mail quality and allergic reactions for some of the more sensitive mail recipients. Early improvements in mail quality were achieved through NIST-improved packing guidelines. NIST also revised irradiation settings through an additional series of on-site tests using the NIST alanine dosimetry system. However, since the mail from the Brentwood and Trenton mail facilities were double-bagged because of their contamination level, the recommendation to slit these bags open for venting immediately after irradiation had one of the most profound effects on the mail quality. Large packages were the last to be sanitized. For this application, highly penetrating bremsstrahlung (high-energy x-ray) beams are required to treat the wide range of package dimensions expected. Totes were constructed (102 cm long, 61 cm wide, 91 cm high) to contain the packages and tests were conducted with surrogate packages. This process is slower and requires more passes through the beam; however, the resulting dose distribution is far more uniform and the mail quality is much improved.

This cooperative effort between NIST, Ion Beam Applications (IBA) and the USPS led to protocols for the processing of parcels with high-energy x rays (from electron beam conversion). About this same time, a team of Federal government and industry representatives drafted a documentary standard that set requirements for validation and routine control of the decontamination process.

In the first year, $\approx$ 3 million articles of contaminated mail were sanitized and safely delivered to their destination (the total for all mail irradiated in the first year is $\approx$ 70 million). Some Federal government mail (defined by zip code) continues to be treated with ionizing radiation. At the end of 2003, about 4000 tons of letter mail and 200 tons of parcels had been sanitized since the process began late in 2001.

2. Radiation Treatment of Passenger Luggage

2.1. Threats to Agriculture

The agricultural industry continues to guard against foreign pests that threaten severe economic consequences. The U.S. Animal and Plant Health Inspection Service (APHIS) defends against this threat every day at more than 80 international airports throughout the United States. Approximately 100 million passengers carry 150 million articles of luggage through these ports each year. About 30 % of this luggage is categorized as high risk. Inspecting upwards of 50 million articles of luggage is a formidable task. The new threats posed by terrorism have raised the level of concern for APHIS inspectors. As increasing the number of inspectors is difficult due to budget constraints, APHIS is considering a technological solution to mitigate these threats.

2.2. Feasibility Study

NIST has a study in progress to examine the technical issues and barriers to the use of irradiation to mitigate common bioterrorism agents and insects in high-risk passenger luggage. The attractive features of this solution are: the individual pieces of luggage do not have to be physically opened and inspected; bioterrorism agents that are concealed, or not easily identified by an inspector, can be treated; the risk of contaminating inspectors or facilities using this treatment method is very low; and, radiation doses can be selected to neutralize a variety of bioterrorism agents, diseases and insects.

Most concerns regarding the radiation sensitivity of luggage contents are not at issue because a large number of common items (*e.g.*, food) are prohibited. However, care should be taken to minimize the absorbed dose to luggage and its contents as not to destroy or render them unusable. Another consideration is throughput; the irradiation equipment must be capable of processing luggage at a rate that does not significantly delay passengers. The purpose of this study is to develop irradiation specifications, procedures, and protocols that will ensure that broad classes of bioterrorism agents in passenger luggage will be neutralized without damaging luggage contents and inconveniencing passengers with long delays.

Table 1. The published D_{10} values for selected microorganisms.

Organism	D_{10} Value (kGy)	Substrate	Temp. (°C)	Ref.
Campylobacter jejuni	0.16 - 0.20	Poultry	5	[1]
Escherichia coli O157:H7	0.30	Beef/Poultry	5	[2,3]
Listeria monocytogenes	0.45	Beef	5	[3]
Salmonella spp.	0.70	Beef	5	[3]
Staphylococcus aureus	0.46	Beef	5	[3]
Bacillus cereus spore	2.46	Beef	5	[4]
Clostridium botulinum	3.56	Poultry	-30	[5]
Moraxella nonliquifaciens	5.3 - 5.8	Beef	-30	[6]
Bacillus anthracis	5.5		ambient	[7]
Foot & mouth disease	5.3		-68	[8]
Swine vesicular disease	1.1 - 5.2		ambient	[9]
Vesicular stomatitis	2.9		-68	[8]
Newcastle	5.2		ambient	[9]
Rift Valley fever	< 2.0		ambient	[10]
Bluetongue	8.3		-68	[8]
Sheep/goat pox	2.2		ambient	[10]
Hog Cholera	5.5		ambient	[8]
Influenza A	2.5 - 7.1		-40	[9,11]

To date, the work has focused on the assembly of critical input data and the development of a coupled computational-experimental verification approach for estimating the radiation dose that can be delivered to passenger luggage.

2.3. Bioagent Data

The list of agents considered a threat to U.S. agriculture and commerce was assembled and a list of their D_{10} values is presented in Table 1. The D_{10} value is the radiation dose required to reduce the population of bacterial pathogens by 90 %.

2.4. Luggage Data

Typical airline luggage restrictions for ordinary handling are a total linear length (length + height + width) $\leq$ 1.57 m (62 in.) and a total weight $\leq$ 31.8 kg (70 lbs). However, virtually any size and weight can be transported on board. Luggage dimensions were collected from sales information of well-known luggage merchandisers. Information on a total of 138 models/sizes of luggage was collected, including type, construction, and dimensions. The collection ranges from small beauty cases, through roll-on cases, garment bags, duffels, wardrobes, up to foot lockers and steamer trunks. This information serves as a guide to available luggage, their dimensions, and their compositions. The luggage volumes ranged from 0.02 m^3 to 0.28 m^3. These sizes require the use of the more penetrating bremsstrahlung (high-energy x-ray) beams. The size of a carrier (tote) used to transport the luggage through the radiation beam was compatible with the steel tote used by IBA in their irradiation of U.S. Postal Service parcels.

A representative set of unclaimed airline luggage (with contents) was lent to NIST by the Federal Aviation Administration (FAA) for this study. The sizes cover the typical range, with the possible exception of the extra-large trunks, and are representative in weight and

Table 2. Dosimetry results for tote 2.

Luggage Description	Contents	Density of Contents (g/cm^3)	Dimension Along Beam (cm)	Absorbed Dose (kGy)
hard shell	copy paper	0.75	21.0	4.7 - 5.0
hard plastic	cardboard	0.19	44.5	5.8 - 6.0
particle-board footlocker	cardboard	0.16	36.5	5.6 - 5.7
soft polyester	plastic clips	0.26	48.3	4.8 - 4.9

construction material extracted from the FAA database. The densities range from 0.11 g/cm^3 to 0.33 g/cm^3, with a mean of 0.16 g/cm^3. A container filled with only close-packed books could have a density up to ≈ 0.75 g/cm^3, and we plan to assume this value for the highest density expected for benign contents of luggage in future calculations.

2.5. Test Runs at 5 kGy

Four test totes were irradiated at the IBA facility, using their 5 MeV x-ray beam for a target test dose of 5 kGy. The test included ≈ 200 dosimeters along with temperature sensors placed in the tote/luggage filler material. The beam-processing parameters were 5 MeV beam energy and 25 mA beam current; four passes were used to achieve a 5 kGy dose. All temperature sensors failed to register the minimum value on the detector (37 °C); data diskettes and CDROMs included in the luggage were readable after irradiation; and there was no obvious damage to luggage or contents. The dosimeters all measured between 4.8 kGy and 6.9 kGy, indicating that reasonably small variations in absorbed dose can be expected for typical luggage irradiated by such a beam.

The intent of the tote 1 was to test the ability to model a pure homogeneous container using a homogeneous product with a density and composition close to that of average passenger luggage. The tote was completely filled with large sheets of corrugated cardboard such that the x-ray beam is normal to the large face. Alanine-film dosimeters were placed along three perpendicular axes with the origin at the center of the tote. Dosimeters were placed at 5 cm intervals along the centerline in the beam direction (z axis). On the plane perpendicular to the beam axis in the middle of the tote, dosimeters were placed at 10 cm intervals from the z axis to the cardboard-pad edges to confirm beam and dose uniformity. Additional dosimeters were placed on the front and back cardboard pads near their corners to probe possible dose depression due to edge effects. In the horizontal conveyor (belt) direction, the dose distribution is relatively flat and symmetric, with some increase at the edges presumably due to in-scatter from the steel tote walls. Along the vertical scan dimension, the fairly symmetric dose distribution peaks in the center and droops at the top and bottom presumably due essentially to out-scatter of the photons. The doses in the corners of the central plane are in agreement with the top and bottom doses.

Tote 2 was completely filled with luggage; the contents of each bag included corrugated cardboard, copy paper, or plastic clips. The bags were constructed from a variety of materials, and dosimeters were located in the center of each bag.

Tote 3 was a real-world scenario of bags containing a variety of materials and personal contents. This tested the feasibility of sanitizing in the mode where there are multiple bags deep along the beam direction by measuring the uniformity of dose from bag to bag. Each bag was completely filled with a variety of clothing and other common personal contents that included diskettes, film in a lead pouch, *etc*. Each bag included centrally located do-

Table 3. Dosimetry results for tote 3.

Description	Average Density (g/cm^3)	Absorbed Dose (kGy)
soft vinyl	0.10	6.1 - 6.3
soft polyester and vinyl carry on	0.11	5.5 - 5.9
soft polyester day pack	0.09	5.6 - 5.8
hard plastic	0.18	5.3 - 5.9
soft vinyl	0.11	5.8 - 5.9
cloth (no frame)	0.15	5.2 - 5.8
leather	0.19	5.2 - 5.9
soft vinyl carry on	0.22	5.6 - 5.9

simeters. Also, the possible radiation-shielding effects of high-*Z* materials was tested by including an assembly of two 3 mm thick steel plates, placed perpendicular to the beam direction, sandwiching a plastic plate into which a vial of alanine-pellet dosimeters was inserted.

The absorbed dose inside the lead-pouch film bag was 5.5 kGy; the doses measured on either side of the bag exterior were 5.5 kGy and 5.7 kGy. The absorbed dose between the steel plates was 5.2 kGy; the doses measured on either exterior side of the steel plates were 5.0 kGy and 5.6 kGy.

The purpose of tote 4 was to measure the dose obtained in a single-bag irradiation configuration with its short dimension along the beam direction, as a possible real-world scenario in which each bag is placed individually on a moving conveyor. The measured doses ranged from 6.4 kGy to 6.9 kGy.

2.6. Test Runs at 25 kGy

Test irradiations were done on five totes at IBA with a target minimum dose of 25 kGy. The beam-processing parameters conditions were 5 MeV beam energy and 25 mA beam current. Tote 1 served as a reference and was filled with a homogeneous product; as in the 5 kGy runs, it was completely filled with large sheets of corrugated cardboard such that the x-ray beam is normal to the large face. The dosimetry confirmed that the depth-dose patterns were comparable to the 5 kGy test.

Tote 2 included three pieces of newly purchased luggage to assess radiation damage at 25 kGy. The balance of the tote was filled with used luggage. Various experiments were included in separate bags. One bag included a "steel sandwich" experiment identical to that of tote 3 in the 5 kGy run described in the preceding section, except that the steel thickness was increased to 1.27 cm; the remainder of the bag was filled with bubble wrap. The dosimetry and computer modeling for this test is shown in Figure 3.

Tote 3 was used to determine the dose delivered inside a lead-shielded container. A lead box with 1 cm thickness was placed inside a trunk filled with cardboard. The lead box contained cardboard and alanine dosimeters. Dosimeters were also placed outside the lead box along the beam direction. The dosimetry results are shown in Figure 4.

Tote 4 was designed to measure the dose distribution in a high-density product, namely, copy paper at 0.75 g/cm^3. Dosimeters were placed between packages of paper, each containing 500 sheets, to measure the distribution of absorbed dose along the beam direction. These data are shown in Figure 5.

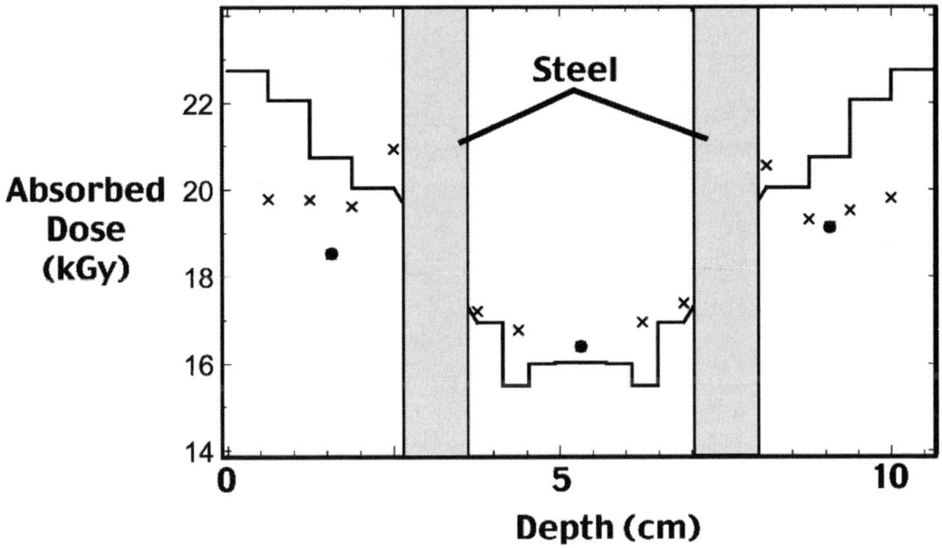

Figure 3. Depth-dose distribution through steel plates: Monte Carlo calculation (–); alanine film dosimetry (x); and alanine pellet dosimetry (•).

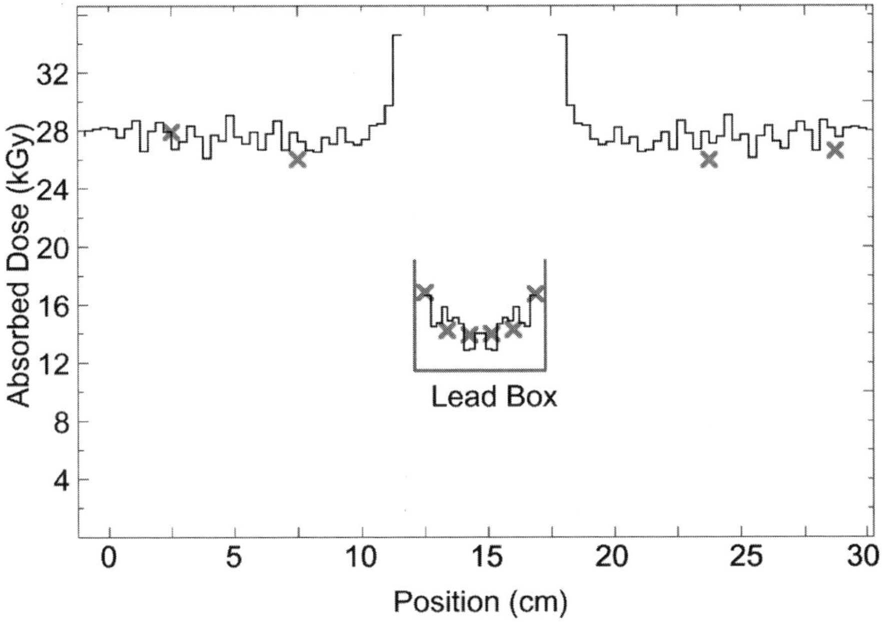

Figure 4. Depth-dose distribution through a lead box: Monte Carlo calculation (–); alanine dosimetry (x).

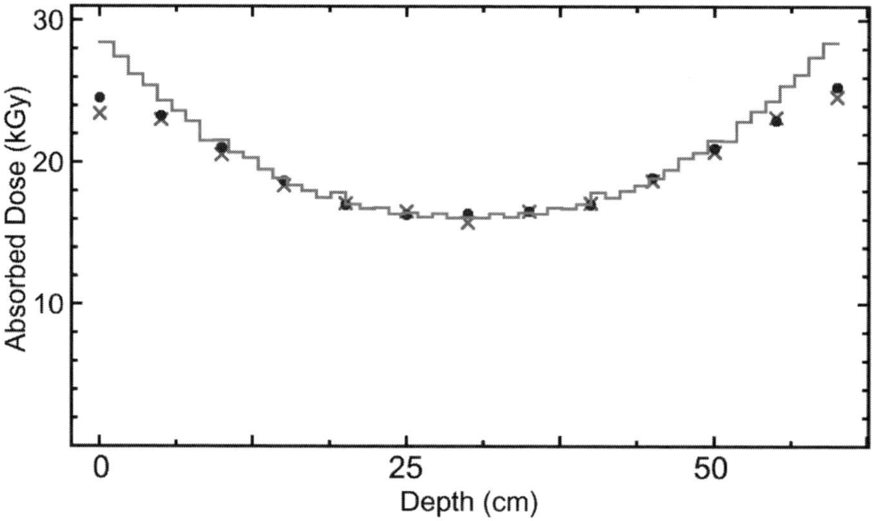

Figure 5. Depth-dose distribution through totes filled with paper: Monte Carlo calculation (–); alanine dosimetry in the middle tote (x) and end tote (•).

The measured maximum temperatures for the different totes ranged from 40.0 °C to 48.9 °C. Other observations included: no noticeable structural damage to luggage (new or used); odors from radiolytic products and ozone were similar to that of irradiated mail; yellowing of plastic in blister pack containing batteries (batteries remained functional); CDROMs and diskettes were undamaged and readable.

2.7. Monte Carlo Simulations

The specifics of the IBA TT300 Rhodotron (proprietary information) were obtained from IBA to aid the Monte Carlo calculations. The three-dimensional Monte Carlo calculations have been developed by altering MCNP4C code to obtain photon phase-space data of the bremsstrahlung output. These phase-space data are used as the source term for calculations of the subsequent transport through the uniform-cardboard tote assembly (and later through any complex configuration) by an altered version of the ACCEPT/ITS3 code. The three-dimensional Monte Carlo results were obtained in two steps. In the first step, calculations were done for a simplified model of the bremsstrahlung source using the Monte Carlo code MCNP. MCNP is a three-dimensional Monte Carlo transport code for electrons, photons and neutrons, the first two particle types being relevant to this work. The necessary output of this run was a description of the phase space of the particles exiting the bremsstrahlung converter. The phase-space results were obtained for the static case of a beam of 5 MeV (and, later, 7 MeV) electrons of a fixed spot size incident normally on the converter. The bremsstrahlung output generated using MCNP was later checked against a simulation using the PENELOPE code. Minor differences were observed at low photon energies. The second part of the calculation is the actual modeling of the transport of the bremsstrahlung photons through the totes. The desired output was the dose to the product contained within the totes. The calculation is naturally split in this way, as the generation of the phase space is quite time consuming relative to the dose calculation, and the same phase-space description can be used regardless of what fills the totes. In this second step, the MCNP code could have been utilized. However, due to some well-known inadequacies of this version of the code in the treatment of problems with many transport zones, the ACCEPT code from the Integrated Tiger Series (ITS) was used in this second part. A code was written to process the MCNP output into a form that could be utilized by ACCEPT. The ACCEPT code was modified to admit a phase-space source and to scan that source along the directions of the scan horn as actually occurs in the Rhodotron. The motion of the totes along the conveyor was simulated by scanning the source along the direction of the conveyor, enabling a realistic simulation of this dynamic system. Wide-angle bremsstrahlung emission from the source was discarded in accord with the physical collimation utilized by IBA. The actual product configuration was faithfully modeled as six totes, stacked two-high along the direction of the scan horn and three-wide along the direction of the conveyor belt. The dose was obtained *versus* depth into the cardboard for the center two totes, the outer four totes serving to realistically model the attenuation and scatter of the photons emitted at large angles. Often times, in modeling a system, some details of the configuration are obscure or missing. This usually makes it necessary to normalize to experiment in one case. In our investigations, enough detail was available to avoid this normalization step.

Still, the Monte Carlo simulations can not be expected to exactly match the measured doses, due variously to imperfect knowledge of the actual details and performance of the accelerator system and of the tote and contents, as well as to uncertainty in the measured results. Figure 5 is a comparison of the experimental data with the Monte Carlo calculations for a tote, where the distribution of absorbed dose is in the direction of the beam. Clearly, the calculated and experimental results show some differences; the calculated results show a rise at the interface with the tote walls with respect to the center of the totes whereas the experimental data are more uniform. However, these results demonstrate that the simulations can serve as a reasonable guide and can confidently be used as a predictive tool in the absence of measurements.

3. Other Opportunities

3.1. Imports

Ship containers entering the country through several major U.S. ports pose a significant smuggling risk. A means to x-ray these containers to examine their contents that would not significantly reduce the throughput would aid security efforts. Because U.S. inspection agencies can set container guidelines to suit the inspection technology; all options are open to facilitate this type of inspection service.

3.2. Military and Law-Enforcement Applications

U.S. military troops face the threat of chemical- and biological-agent attacks throughout the world. There is a possibility that military materials might need to be decontaminated. Here, field-deployed portable electron-beam accelerators are a reasonable solution. Along these same lines, the decontamination of physical evidence should be considered; this application would apply to national and international law-enforcement agencies as well. The use of ionizing radiation to decontaminate materials questions the applicability of standard forensic tests used in law enforcement. As a first test, the Ionizing Radiation Division and Biotechnology Division of NIST collaborated to show that a high dose of ionizing radiation does not interfere with standard DNA profiling tests [12].

3.3. Basic Research

Volatile organic irritants and the degradation of paper resulting from the mail irradiation process underscore the need for radiation-effects studies on common materials for safety and archival efforts. The radiation effects on certain sensitive materials such as magnetic media and their packaging material should also be examined.

Another issue is the consideration to improve irradiator throughput by raising the photon energy from 5 MeV to 7 MeV; a risk assessment for induced activity should be undertaken.

3.4. Critical Data

The research described in this paper relies heavily on the quality of published D_{10} values. A reevaluation of these data by a team of recognized experts would be welcomed by the research community. A committee of plant- and animal-pathogen experts should be convened and tasked with assembling and evaluating the available data with the intent of producing a report that summarizes their consensus opinion.

Acknowledgements

The NIST Ionizing Radiation Division acknowledges the support (CB-8902-NISTIRD-91240) of the Technical Support Working Group (TSWG) and the cooperation of the Federal Aviation Administration for the luggage irradiation project, as well as the Titan Corporation and the Ion Beam Applications Corporation for use of their facilities.

References

[1] Lambert and Maxcy (1984) J. Food Sci. 49:665.
[2] Clavero et al. (1964) Appl. Environ. Microbiol. 60:2069.
[3] Thayer et al., (1995) J. Food Sci. 60:63.
[4] Thayer and Boyd (1994) J. Food Prot. 57:758.
[5] Anellis et al., (1977) Appl. Environ. Microbiol. 34:823.
[6] Maxcy et al., (1976) Tech. Rep. 76-43 FSL U.S. Army Natick R&D Command.
[7] Bowen et al., (1996) Salisbury Medical Bulletin 87; Sup P: 70.
[8] House et al., (1990) Can. J. Micro. 36:737.
[9] Sullivan et al., (1971) Appl. Micro 22:61.
[10] Thomas et al., (1981) Can. J. Micro. 45:397.
[11] Lowy et al., (2001) Antiviral Res. 52:261.
[12] Desrosiers (2004) Radiat. Phys. Chem. 71:479.

Research Directions at State Research Center of Virology and Biotechnology VECTOR. International Collaboration is an Efficient Option for Infectious Disease Control and Combating Bioterrorism

Raisa A. MARTYNYUK and Lev S. SANDAKHCHIEV
State Research Center of Virology and Biotechnology VECTOR,
Russian Federation Ministry of Health
Koltsovo 630559, Novosibirsk Region, Russia

Abstract. State Research Center of Virology and Biotechnology VECTOR is one of Russia's largest scientific research and production facilities whose major activities are focused on carrying out basic and applied research in a wide area of natural sciences, development and manufacture of therapeutic, preventive, and diagnostic products for public health and agriculture.

The State Research Center of Virology and Biotechnology VECTOR that is an institution of the Russian Ministry of Health. Before 1991, VECTOR was involved in biodefense research.

Today, the State Research Center of Virology and Biotechnology VECTOR is one of Russia's largest scientific research and production facilities whose major activities are focused on carrying out basic and applied research in a wide area of natural sciences, development and manufacture of therapeutic, preventive, and diagnostic products for public health and agriculture.

Currently, VECTOR is a large research and production complex that is comprised of several research institutes, production companies, and other departments (Organization's Chart). The main asset for the Center is unique research and experimental capabilities that allow research on extremely human-and animal pathogenic viruses whereas total biosafety for the personnel involved and the environment is guaranteed. The well-trained scientists and technicians are very experienced in handling extremely hazardous viruses. VECTOR employs a total over 1800 scientists and technicians. Of these, approx. 900 are directly involved in conducting scientific research. Over 170 VECTOR employees hold DSci and Ph.D. degrees.

Before 1992, VECTOR received all of its funding from the federal budget and just began to establish manufacturing activities.

In 1989, it became quite clear to us that VECTOR should undergo a massive restructuring effort to adapt itself to the changing economic conditions and a significant cutback in federal research budget funding. In order to address this challenge, we devised a program of VECTOR's long-term development, with a particular emphasis on conducting research on infectious diseases of importance to public health such as HIV, tick-borne encephalitis, viral hepatitides A and B, measles, influenza etc., and development of diagnostic tests, vaccines and antivirals as well as a set up of manufacture of diagnostic, therapeutic and prophylactic products. In 1993, VECTOR became a State Research Center and started to receive federal budget funding to support its R&D activities, through government civilian programs. The development of pharmaceutical manufacturing activities was supported by government investments and credits, which allowed us to renovate and upgrade several facilities and purchase necessary equipment.

During these years, dramatic changes in VECTOR's income pattern have occurred, i.e. in 1999 78 percent of funds came from federal government budget, and in 2003 72 percent of the total incomes was generated by sales of products.

Regarding the importance of international assistance provided under nonproliferation programs, since 1995, VECTOR has been involved in collaborative research programs with ISTC, CRDF, DOE-IPP (USA) and in EU programs. Since recently VECTOR has been carrying out the so-called ISTC Partner Projects with US DHHS (BTEP/FETP), USDA Agricultural Research Service, and DOD (DTRA, DARPA). The total financial support provided to VECTOR under the nonproliferation programs since 1995 has amounted to over US$22.0 million, including over US$19.0 million provided via the ISTC. During this period, VECTOR has had 87 international projects (both completed and active), including 60 ISTC projects. It allowed VECTOR to significantly accelerate the implementation of the restructuring program in terms obtaining important scientific results, improvement of its telecommunication infrastructure, and meeting current biosafety and physical security requirements supporting the kind of pathogen research we do.

Doing international projects has contributed to the implementation of a quality system at VECTOR; the establishment of a certified Institutional Review Board (Ethics Committee) and an Institutional Animal Care and Use Committee (Bioethics Committee); and ensuring compliance of at least part of experimental studies with the GLP requirements. VECTOR scientists are now well integrated into the world scientific community.

In this paper we would like to address the need for international cooperation on combating bioterrorism. During the recent decade, policy makers, military and civilian experts have shown more and more interest in the bioterrorism issue. Possible biological agents of viral or bacterial etiology, scenarios of how to prevent from and respond to the use of these agents, and epidemic response capabilities in terms of the availability of competent personnel, diagnostic and therapeutic products have been discussed and analyzed.

As a rule, the scenarios of bioterrorism incidents are far from being optimistic in terms of both human casualties and costs associated with containing direct consequences of such actions as well as with an economic breakdown in the region affected and lasting psychological effect they produce on the population [1, 2].

Terrorism now is a growth industry and the possibility of a chemical or bioterrorism attack is increasingly defined as "not if, but when". However, even the USA that has longstanding experience in infectious disease control all around the world only in the year 2000 developed a Biological and Chemical Terrorism: Strategic Plan for Preparedness and Response Recommendations of the CDC Strategic Planning Workgroup [3] that involves coordinated response to and elimination of such events by over 10 agencies. This plan is focused on five major areas:

- preparedness and prevention;
- detection and surveillance;
- diagnosis and characterization of biological and chemical agents;
- response; and
- communication.

All these areas propose personnel training, investigation of and total preparedness for detection and elimination of consequences of the possible attack using chemical or biological agents in all the states and cities. The key point is to design a multi-level laboratory network to efficiently warn public health authorities at community, state, district and city levels about biological and chemical agents detected. This plan is aimed at significant reengineering of the existing infrastructure dedicated to infectious disease response and control.

We should especially emphasize important features that make bioterrorism different from other kinds of terrorism [4, 5].

Explosive substances are pretty widespread and not that diverse. Chemical agents that could be used for terrorist purposes are well-studied as potential chemical weapons agents and for many of them detection procedures as well as treatment of those affected and decontamination have been developed. In case of biological agents, it is an absolutely different situation. However, the study of the effect of chemical and radiological factors on the pathogenesis of infectious diseases and investigation into how properties of infectious agents are influenced by these factors appears of enormous importance.

In nature, there is a great variety of viruses, bacteria, and fungi causing diseases in humans, animals or plants. Experts estimate that currently we are aware of far less than 1 percent of existing viruses and several percent of microbes. Nature is continuously creating new pathogens, the so-called "emerging infections", and this potential is just inexhaustible. The most recent example of this is the SARS situation, a disease known as Severe Acute Respiratory Syndrome as well as the avian influenza situation. During the last 20 years alone, over 30 new infectious agents e.g. HIV, Marburg, Ebola, Machupo, Nipah virus and SARS-associated coronavirus have been discovered against which no efficient treatments are available so far.

An outbreak of disease elsewhere on the globe can now be viewed as a threat to any other region of the world. Once an infectious disease, or the insects and animals that carry it, invades a new country or continent, it can prove difficult – if not impossible – to control. This has been the case with West Nile fever, which made its initial appearance on the American continent in 1999 and is now firmly entrenched and spreading [6]. Even high economic development of these countries did not hinder the spread of this infection. According to the Centers for Disease Control and Prevention (CDC), by September 2002 West Nile virus activity had been documented in 42 states and the District of Columbia: 1 460 human cases of the West Nile virus were reported, with 66 deaths [7]. Initial costs associated with cases of West Nile fever in New York were placed at almost $100 million [6]. Health Canada reported 17 suspected cases of West Nile, 3 confirmed cases, including 1 death [8].

Known diseases such as influenza, TB, malaria and some others, through their changeability, can relatively easily overcome conventional immunization and drug-based approaches to prevention and therapy.

The human kind has been fighting a biological war against microbes since its emergence and even now, according to WHO, infectious diseases account for 24.7 percent of fatality worldwide. In developing countries with public health underfunded, this figure increases to 45 percent. Infantile mortality from infectious diseases reaches 63 percent of all infantile deaths and 48 percent of untimely deaths (under the age of 45) are brought up by infections [9, 10, 6].

Though biological weapons-and bioterrorism experts often operate with a limited list of several dozens of infectious agents on it, we should not underestimate the possible terrorist use of any of diverse pathogens existing in nature.

So the task of establishing a global system of surveillance of possible natural or artificial outbreaks is far more difficult than that of chemical agents or explosives.

It is important to realize that biological agents act in time, have a latent period during which the carrier of infection may find herself/himself in another city or even country, where the outbreak of disease may be actually identified, and it may take much time to prove the bioterrorist use of microorganisms since it will require a comprehensive epidemiological analysis, e.g. investigation of all the stages of manufacture and distribution of food-stuffs in case of food poisoning.

A qualitatively more difficult problem is that of keeping the inventory of the pathogens during research work since during such activities the biological agents, as a rule, grow in quantity and can be represented by not only individual pathogens but also by being present in experiment in the form of infected cell cultures, laboratory animals etc. Insignificant, hardly accountable quantities of a biological agent, may pose a real threat in terms of uncontrolled leakage of biological material. Unfortunately, this problem does not yet have either an engineering or technical solution. In fact, it is determined by the human factor, i.e. it is necessary to adopt criteria and requirements to personnel to be allowed to work with pathogens, even within the highly secured laboratory facilities.

An associated problem is that highly pathogenic agents are many and they might be accessed during natural outbreaks of disease. Moreover, they can be engineered through simple laboratory manipulations on the non-pathogenic microorganisms available. The well-known case of the terrorist use of Salmonella in a salad bar in Oregon in 1984 resulted in sickening over 700 individuals. However, first it was regarded to be a natural outbreak and only one year later it was proven that *Salmonella* was used by religious cult extremists to prevent voting in Oregon. By the way, the US public learned about that many years after.

Therefore, it is medical staff that turn out to be the first to have to deal with biological incidents and it is the public health capabilities that determine the preparedness of a country, region or city for a timely detection and elimination of consequences of the use of biological agents. Therefore, financial and organizational efforts should be focused on civilian rather than military agencies.

The nation must be prepared to deal with detection and elimination of consequences of outbreaks caused by any biological agents, including both conventional and exotic species of microorganisms. The existing national systems of nation-wide epidemiological surveillance and control of infectious diseases should be capable of identifying, containing and eliminating an infectious disease outbreak regardless of whether it is the result of natural manifestation of a pathogen or its deliberate use.

I should point out these features of control over biological agents and say that that international collaboration in this area is both extremely important and urgent in order to set up a system of efficient alert and response. This issue was specifically addressed in May 2001 during the 54th World Health Assembly in the report by the Secretariat "Global Health Security - Epidemic Alert and Response" (http://www.who.int/gb/EB_WHA/PDF/WHA54/ea54r14.pdf). It was noted that in 1995 the World Health Assembly adopted resolutions WHA48.13 on new, emerging and re-emerging infectious diseases and WHA48.7 on the revision and updating of the International Health Regulations. The WHO totally realized the need for enhancing epidemiological and laboratory surveillance at national level as "the main defense against the international spread of communicable diseases".

The WHO Secretariat pointed out the increased possibility of intentional use of agents causing infectious diseases and emphasized those natural epidemics and those due to the deliberate use of biological agents may manifest themselves in the same manner.

In 1997, WHO established a special system to seek, collect and verify information on reported outbreaks based on close cooperation of WHO Collaborating Centers with governmental and nongovernmental agencies, which is available as confirmed disease outbreak news on the WHO web site (www.who.int/disease-outbreak-news/) and in the WHO Weekly Epidemiological Record (www.who.int/wer). At global level, laboratory networking takes place (http://www.who.int/csr/en/), focusing on such infections as hemorrhagic fevers (including Ebola virus), poliovirus; preparation of databases such as the WHO antimicrobial resistance data bank (ARInfoBank) (www.who.int/emc/amr.html), influenza

FluNet (http://oms2.b3e.jussieu.fr/flunet/), rabies RabNet (www.who.int/emc/diseases/zoo/rabies.html), and some others. WHO also called the member states to establish partnerships to involve both civilian public health and military medical capabilities.

WHO continuously draws the attention of its member states to the ultimate role of national potential to ensure epidemiological welfare of other countries and so it plans to expand national training programs in intervention epidemiology worldwide, and the Training in Epidemiology and Public Health Interventions Network (TEPHINET). Major conclusions based on discussions of the Secretariat Report were reflected in resolution WHA54.14 "Global Health Security: Epidemic Alert and Response" (http://www.who.int/gb/EB WHA/PDF/WHA54/ca54r14.pdf). A good example that deserves serious attention is the establishment of the WHO Bureau in Lyon (France) as a model for using national potential to contribute to training of personnel for countries at high epidemic risk (http://www.who.int/infectious-disease-news/newsletter/vol2-6September-ctober2001/vol2-6-eng.pdf).

At global level, huge resources to combat infectious diseases are available already. Certainly, these will be used to counter bioterrorism incidents [11]. They include hundreds of WHO Collaborating Centers worldwide specializing in certain infections; a Pan-American Health Organization (PAHO) laboratory network; International Clinical Epidemiology Network (INCLEN); the Pasteur Institutes network; international research centers network of the National Institutes of Health (NIH) that involves many universities across the USA; Centers for Disease Control (CDC) in numerous countries many of which conduct epidemiological surveillance and provide field epidemiology training for different regions. US Army and Navy also established a specialized network of research centers in several countries. It should be noted that this particular resource is very much focused on specific tasks and, except for the Epidemiologic Intelligence Service (EIS) centers, is not oriented on detection and identification of the entire pathogen range.

As a matter of fact, to localize and contain unusual outbreaks posing threat to global public health WHO has set up taskforces to be deployed during the life of such outbreaks.

Another approach was proposed by an outstanding epidemiologist Dr. D. A. Henderson [12] who, based on many years of his experience as the leader and actual participant in the global smallpox eradication program, arrived at the conclusion that fixed-site international centers be established in 15 regions of the world. These should include:

- in-and outpatient capabilities to deal with infectious diseases;
- research and diagnostic laboratories;
- epidemiological teams to function like the epidemiological intelligence service (EIS) to cover regions with populations of 2 to 5 million;
- education and training capabilities to provide training to national and international personnel.

Systematic studies of a specific region make it possible to obtain invaluable databases, investigate into different factors that can influence the epidemiological situation, and identify unusual cases requiring careful examination.

To detect and counteract bioterrorism, the National Institute of Allergy and Infectious Diseases (NIAID/NIH) has developed a biodefense research agenda for CDC A Category Pathogens such as smallpox, plague, tularemia, anthrax, botulinum toxin, etc. (http://www.niaid.nih.gov/biodefense/research/biotresearchagenda.pdf). As part of this strategic plan, the NIAID is establishing the Regional Centers of Excellence for Biodefense and Emerging Infectious Diseases Research (RCE) Program and Biocontainment Laboratories (BL) (http://www.niaid.nih.gov/biodefense/rblrce.htm). The RCEs and BLs will contribute to the need for new infrastructure and research resources necessary for identifying and responding to emerging diseases and bioterrorism events. In addition, the centers will:

- Conduct investigator-directed research;

- Train researchers and other personnel for biodefense research activities;
- Develop translational research capacity to test and validate vaccine, therapeutic and diagnostic concepts for emerging infectious diseases, including agents of bioterrorism;
- Establish and maintain core facilities for center investigators from academia and the private sector for the performance of experiments and for the testing and evaluation of vaccines, therapies and diagnostics for NIAID Category A-C priority pathogens; and
- Prepare and make available facilities and scientific support for first-line responders in the event of a national biodefense emergency.

The European Union Commission recently developed a program on Cooperation in the European Union on Preparedness and Response to Biological and Chemical Agent Attacks (Health Security) [13].

NATO has recently revised its science program to address the new challenges in the post September 11th world. The revised science program now has a new name, NATO Program for Security Through Science [14]. The new name reflects more closely the aims and purposes of the revised program. Support will now be offered only for collaboration on security-related Priority Research Topics, which follow the new directions and objectives of the Alliance. The Priority Research Topics are in the following areas:

1. Defense Against Terrorism;
2. Countering Other Threats to Security.

The World Health Organization together with the International Center of Genetic Engineering and Biotechnology (ICGEB) and several nongovernmental organizations - Program for Appropriate Technology in Health (PATH), International Clinical Epidemiology Network (INCLEN) and Training Program in Epidemiology and Public Health Interventions Network (TEPHINET), the so-called Alliance against Infectious Diseases, in the follow-up of the US Institute of Medicine recommendations prepared (in 2000) a program proposal "Global Monitoring, Research and Training to Control Infectious Diseases" (http://www.fas.org/bwc/papers/allaid.htm).

In the initial stage of the program, 10-12 laboratories or institutes would be identified that would locate in strategically important regions at high epidemic risk and with insufficient surveillance capabilities. Those laboratories should have laboratory and clinical study capabilities and a potential for conducting epidemiology work, access to air and ground transport, possibility of telecommunications installation, and prospects for the center's future expansion.

Centers thus identified would have a WHO Collaborating Center's status and preferred access to WHO programs and those of Health Ministries in state-parties, and they would be coordinated by the WHO Office of the Strategic Alliance.

Each center, in its turn, would be established taking into account the region's specific needs and, in the initial stage, provided with necessary resources to create the most advanced potential for diagnostic, clinical and epidemiological activities. It also would be provided with telecommunications to be able to communicate with the other centers as well as regional, federal, and international agencies involved in infectious disease surveillance and response.

Each center would establish regional networks to include clinics, institutes, education establishments and others and it would have intensive participation in the region's infectious disease programs. The regional network would involve enterprises manufacturing specialized pharmaceutical products that, through technology transfer, would be given an opportunity to meet the region's needs for standard diagnostic tests and therapeutic products.

The regional network should also involve research laboratories that develop diagnostic and therapeutic products as well as vaccines; biosafety research laboratories studying the safety of biological substances and microorganisms to humans and the environment.

The program envisages that within 8-10 years a worldwide network of regional centers would be up and running and so a long-term sustainable regional potential for communicable disease control would be created. It is proposed that some of these centers would become the centers of excellence like CDC, NIAID, and ICGEB.

The idea of international/ regional centers of excellence seems to be very promising [15, 16, 17]. Thus, SRC VB VECTOR, with US DHHS Biotechnology Engagement Program's support, is carrying out an ISTC project aimed at developing a concept for the establishment of an International Center for the Study of Emerging and Re-emerging Infectious Diseases (ICERID). By an *International Center* we mean an international organization established by an intergovernmental agreement, similar to those of ISTC or the Joint Institute for Nuclear Research in Dubna, CERN in Switzerland or International Center for Genetic Engineering and Biotechnology in Trieste (Italy). Though the process of establishing the *International Center* is complex and may take several years to complete, the proposed arrangement would provide for a long-term strategic collaboration, which is far less subject to political or economic conjuncture fluctuations in Member States. International partnership would accelerate and facilitate the study of dangerous pathogens and the development of state-of the-art products for diagnosis, prophylaxis and therapy for public health and countering bioterrorism. It is, however, important to establish an appropriate regimen of use of infectious agents and scientific results obtained to avoid their possible misuse for illicit purposes.

A Global Partnership, launched by the G8 leaders at the summit at Kananaskis in Canada in June 2002 appears to be a qualitatively new stage of international collaboration on nonproliferation and countering bioterrorism. (http://www.sgpproject.org/jointstatement.html; http://www.csis.org/pubs/2003_protecting1.pdf). It seeks to achieve the following goals:

- promote multilateral treaties that help prevent the spread of weapons, materials and know-how;
- account for and secure these items;
- promote physical protection of facilities;
- help detect, deter, and interdict illicit trafficking;
- promote national export and transshipment controls; and
- manage and dispose of nuclear, biological and chemical weapons materials.

During the June 2003 summit at Evian, France, the leaders of countries, members to the Global Partnership, discussed some first outcomes and the nearest future action items in order to "prevent terrorists, or those that harbor them, from acquiring or developing nuclear, chemical, radiological and biological weapons; missiles; and related materials, equipment and technology." Together, they pledged to raise $20 billion over 10 years to support these goals. It should be noted that apart from the above technical goals that the Global Partnership seeks to achieve, there is an increasing trend towards broader international research collaboration to develop protections from biological agents that may be used by bioterrorists. This includes development of prevention, diagnostics and treatments as well as detection of possible terrorist attacks involving the use of biological agents.

While assessing very highly the above national and international efforts in the nonproliferation area, in strengthening confidence and transparency and countering bioterrorism, we would like to offer several recommendations for the action agenda:

1. In developing plans to counter biological and chemical terrorism, these should be based on the World Health Organization's recommendations on enhancing national

preparedness and efficient response. An alert about possible deliberate acts should come from national security sectors whereas the responsibility for responding should lie with the public health sector as well as with veterinary, food safety and supplied water quality sectors.

2. There is an urgent need for long-term (at least 5 years) international programs to be launched on basic and applied research on the poorly known pathogens affecting humans, animals and plants, including assessment of the possible deliberate creation and use of genetically modified microorganisms. Special attention should be paid to the study of diseases with the potential to cause epidemics such as smallpox, Ebola fever, SARS, etc.
3. To ensure efficient implementation of the scientific research programs, a simplified set of export controls for strains and related information should be adopted. At the same time, we should strengthen the monitoring over the possible "drain" of biological material and technology from the Global Coalition's research sector. The "human factor" should be paid special attention to. Unified requirements to storage, inventory of and work on pathogens should be adopted as well as unified biosafety and physical security standards supporting this type of research.
4. To implement the research programs and ensure preparedness for natural and deliberate outbreaks, it is proposed that regional international Centers of Excellence for Bioterrorism and Emerging Diseases Research be established. It is important, however, that scientists from member-states to such centers be provided on-site access to conduct joint studies at these centers. As a starting point for discussing this proposal, data produced by the Alliance against Infectious Diseases could be used.

I wish to acknowledge the key role played by the staff of the Russian Federation Ministry of Industry, Science and Technologies, Russian Federation Ministry of Health, RAO BIOPREPARAT, Russian Federation Ministry of Foreign Affairs, Russian Academy of Sciences, Russian Academy of Medical Sciences, Institute of International Security of the Russian Academy of Sciences, those of the U.S. Department of State, ISTC, Cooperative Threat Reduction Program (DTRA/CTR), the U.S. Department of Energy, U.S. DHHS, CDC, NIH, U.S. NAS, USDA, CRDF in promoting US-Russian collaboration.

References

[1] R. Preston. The Bioweaponeers. The New Yorker, March 9, 1998, p. 52-65.

[2] R. Preston. Bio-Warfare-Fiction and Reality, Genetic Engineering News, March 1, 1998, p. 6-39.

[3] April 21, 2000 / 49(RR04); 1-14 Biological and Chemical Terrorism: Strategic Plan for Preparedness and Response Recommendations of the CDC Strategic Planning Workgroup.

[4] Chemical and Biological Terrorism. Research and Development to Improve Civilian Medical Response. Committee on R&D Needs for Improving Civilian Medical Response to Chemical and Biological Terrorism Incidents. Health Science Policy Program. Institute of Medicine and Board on Environmental Studies and Toxicology. Commission on Life Sciences. NATIONAL RESEARCH COUNCIL. National Academy Press, Washington, D.C. 1999.

[5] Proceedings of the Eleventh Amaldi Conference on Problems of Global Security (November 18-20, 1998, Moscow, Russia). Moscow "Nauka" 1999.

[6] Heymann D.L. Strengthening Global Preparedness for Defense against Infections Disease Threats. WHO, 2001. http://www.who.int/emc/surveill/index.html.

[7] West Nile virus in the United States - Update 3, 2002. http://www.who.int/disease-outbreak- news/n2002/september/17september2002.html.

[8] West Nile virus in Canada - Update 2, 2002. http://www.who.int/disease-outbreak-news/n2002/september/20september2002.html.

[9] Communicable diseases 2000. WHO/CDS/2001. http://www.who.int/infectious-disease-news/CDS2000/index.html.

[10] Anker M., Schaaf D. WHO Report on Global Surveillance of Epidemic-prone Infectious diseases //WHO/CDS/CSR/ISR/2000.1.

[11] Emerging infections. Microbiological threats to health in the United States, J. Lederberg, R.E. Shope and S.C. Daks, Jr., Ed. National Academy Press, Washington, D.C., 1992.
[12] Emerging viruses, S.S. Morse, Ed., Oxford University Press, 1993.
[13] http://europa.eu.int/eur-lex/en/com/cnc/2003/com2003_0320en01.pdf.
[14] http://www.nato.int/science/news/2003/n031111a.htm;
http://www.nato.int/science/topics/priority_research_topics.htm
[15] Lev S. Sandakhchiev and Sergei V. Netesov. Strengthening the BTWC through R&D Restructuring: The case of the State Research Center of Virology and Biotechnology "Vector". The Role of Biotechnology in Countering BTW Agents. 2001 Kluwer Academic Publishers. Printed in the Netherlands.
[16] S. Netesov, L. Sandakhchiev. The Development of a Network of International Centers to Combat Infectious Diseases and Bioterrorism Threats. The ASA Newsletter, February 19, 1999, Issue Number 70, PP. 2-6.
[17] Lev S. Sandakhchiev. The Need for International Cooperation to Provide Transparency and to Strengthen the BTWC. E. Geissler et al. (eds.) Conversion of Former BTW Facilities, Kluwer Academic Publishers. Printed in the Netherlands, 1998, PP. 149-156.

Radiation Inactivation of Bioterrorism Agents
L.G. Gazsó and C.C. Ponta (Eds.)
IOS Press, 2005

Differential PCR Diagnostic of Orthopoxviruses

Sergei N. SHCHELKUNOV and Lev S. SANDAKHCHIEV
State Research Center of Virology and Biotechnology VECTOR,
Russian Federation Ministry of Health
Koltsovo 630559, Novosibirsk Region, Russia

Abstract. Methods for rapid identification of orthopoxviruses pathogenic for humans using a RFLP, PCR, multiplex PCR, real-time PCR, and microarray assays are described.

Introduction

The genus *Orthopoxvirus* of the family Poxviridae includes the species pathogenic to humans, such as variola (VARV), monkeypox (MPXV), and cowpox (CPXV) viruses.

VARV causes smallpox and is an exclusively anthroponotic agent. This is the first and yet single infectious disease that was eradicated due to the international medical program under the aegis of WHO [1]. Now, VARV is regarded as a potential bioterrorism agent [2, 3].

Rodents are the natural reservoir of MPXV. Human monkeypox resembles the clinical course of smallpox that was prevalent on the African continent, and is recorded predominantly in Central and Western Africa [1, 4]. The lethal cases as well as human-to-human transfer were recorded mainly within the unvaccinated cohort [4, 5].

CPXV displays the widest host range among the orthopoxviruses. Generally, human cowpox is a benign disease manifesting itself by isolated local lesions [6-8]. In the case of immunocompromised persons, the disease may have a generalized form with lethal outcome [9]. Human cowpox is recorded in the majority of European countries and several countries of Asia and Latin America. Rodents (the main natural reservoir) or home pets and cattle (bridging hosts) represent the main sources of human CPXV infection [6].

Cessation of vaccination against smallpox since 1980 resulted gradually in formation of a large population cohort susceptible not only to VARV, but also to other orthopoxviruses. This formed an opportunity for ever increasing distribution of previously relatively mild infections of monkeypox- and cowpox-types in the human population [4-6]. In particular, outbreaks of human cowpox in Brazil (ProMed, Archive number 20030111.0095) and human monkeypox in the USA in 2003 were recorded for the first time [10].

The potential increase in the degree of danger of orthopoxvirus infections for humans required development of efficient methods for rapid detection and identification of orthopoxviruses pathogenic to humans.

The conventional biological and serological methods used appeared insufficiently effective for rapid diagnostics of orthopoxviruses. The biological analysis takes too long time (3–6 days) and involves handling of special viral pathogens. The serological methods, as a rule, allow only for genus-level identification; moreover, their sensitivities are frequently insufficient for assaying clinical samples.

1. Restriction Fragment Length Polymorphism Analysis of Viral Genomic DNAs

The methods based on genomic analysis may be regarded as an effective and efficient approach to diagnosing viral infections. It was demonstrated [11-14] that restriction enzyme assay of viral DNAs provided a reliable species-level identification of orthopoxviruses. However, this requires propagation of the virus and its purification, demanding specialized equipment, and is time-consuming.

2. Virus Identification Using Polymerase Chain Reaction

The advent of the method of DNA fragment amplification by polymerase chain reaction (PCR) [15] formed the background for designing various techniques appropriate for rapid identification of orthopoxviruses. The method of DNA fragments amplification in PCR makes it possible to recover the specific DNA fragments from trace quantities of genetic material under study, during a short time. It does not require any cultivation of the virus to be tested, which is especially important when we have deal with highly pathogenic strains.

Determination of the genomic nucleotide sequences for a number of strains of several orthopoxviral species [16-22] allowed to reveal species-specific differences of some locuses and develop new methods for virus identification. So far, application of PCR for detection of orthopoxviruses using common pair of oligonucleotide primers to the regions of genes encoding hemagglutinin (*HA*) [23], A-type inclusion protein (*ATI*) [24], and homologue of tumor necrosis factor receptor (*CrmB*) [25] is described. In all these techniques, the DNA fragments obtained by PCR are hydrolyzed with certain restriction endonucleases and separated by electrophoresis; the resulting patterns of subfragments allowed orthopoxviruses to be identified at a species level (Fig. 1).

However, when a large enough set of isolates of an orthopoxvirus species was analyzed, heterogeneity of their restriction fragments patterns became apparent, making interpretation of the results obtained rather ambiguous (Fig. 2) [23-25].

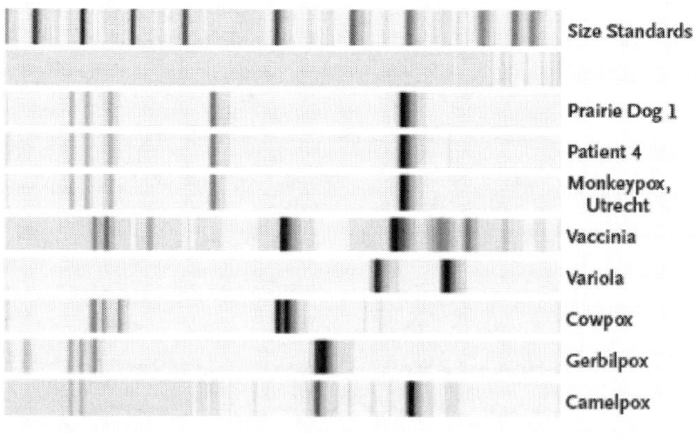

Figure 1. Polymerase-chain-reaction amplification of the orthopoxvirus *HA* gene followed by restriction-fragment–length polymorphism (RFLP). Analysis of samples from patient 4, prairie dog 1, and reference isolates of other orthopoxviruses [14]. *Taq*I was used for the RFLP analysis.

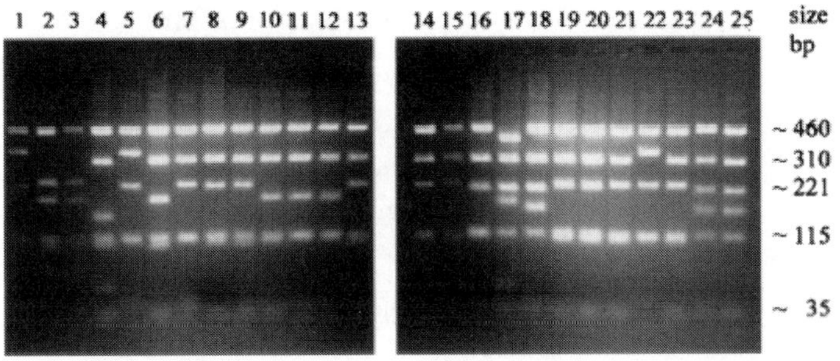

Figure 2. *Nla*III RLFP analysis of PCR-amplified *CrmB* fragments from DNA of CPXV isolates from human and animals. *Nla*III digest fragments after PCR amplification of OPV85 (human) (lane 1), OPV88/L (cat) (lane 2), OPV88/H (cat) (lane 3), OPV89/1 (cat)-M5 (lane 4), OPV89/2 (cat) (lane 5), OPV89/3 (cat) (lane 6), OPV89/4 (cat)-M6 (lane 7), OPV89/5 (cat)-M7 (lane 8), OPV90/1 (cat)-M8 (lane 9), OPV90/2 (human) (lane 10), OPV90/4 (dog) (lane 11), OPV90/5 (cat)-M9 (lane 12), OPV91/1 (cat) (lane 13), OPV91/2 (human) (lane 14), OPV91/3 (cow) (lane 15), CATPOX3 (lane 16), CATPOX5 (lane 17), RAT Moscow (rat) (lane 18), EP-1 (elephant) (lane 19), EP-2 (elephant)-M1 (lane 20), EP-3 (elephant) (lane 21), EP-4 (elephant) (lane 22), EP-5 (elephant) (lane 23), CPV BRT-Atlanta (lane 24), and CPV BRT-Munich (lane 25) are shown [24].

3. Multiplex PCR Analysis

Specific detection of MPXV by PCR was developed using one species-specific oligonucleotide primer pair for gene *ATI* sequence [26]. But it would be very important to use technique which allows to discriminate orthopoxviral species in one step analysis.

Therefore a new method of multiplex PCR assay (MPCR) of orthopoxviruses pathogenic to humans was developed recently [27]. This method displays a high specificity and sensitivity of the analysis in question. The essence of the method designed is that the selected unique oligonucleotide primers allow orthopoxviruses to be identified at a species level in one stage. Four pairs of oligonucleotide primers (three pairs for VARV, MPXV, and CPXV, respectively, and one genus-specific pair) were used in a united polymerase chain reaction producing amplicons of various lengths specific of each orthopoxvirus species in question (Fig. 3).

Used primers allow for discriminating between the VARV subtypes, such as major and minor as well as Central African and Western African MPXV subtypes. The genus-specific pair was used as an internal PCR control for the presence of orthopoxvirus DNA in the sample and discrimination from other genera of poxviruses. Specificity and sensitivity of the method developed were evaluated using DNAs of 57 orthopoxvirus strains, including the DNAs isolated from human case clinical materials (scabs from skin lesions of smallpox human cases infected in 1970–1975 and deposited with the Russian Collection of Variola Virus). Being simple, quick, and exact, the developed MPCR assay allows separate analysis of each sample using each pair of primers to be avoided and orthopoxviruses pathogenic to humans to be identified at a species level in one stage.

4. Real-Time PCR Analysis

When identifying a particular virus species, a real-time PCR is even more efficient because it combines amplification and detection of target DNA in one vessel, thereby eliminating any time-consuming post-PCR procedures, and potentially limiting possible contamination

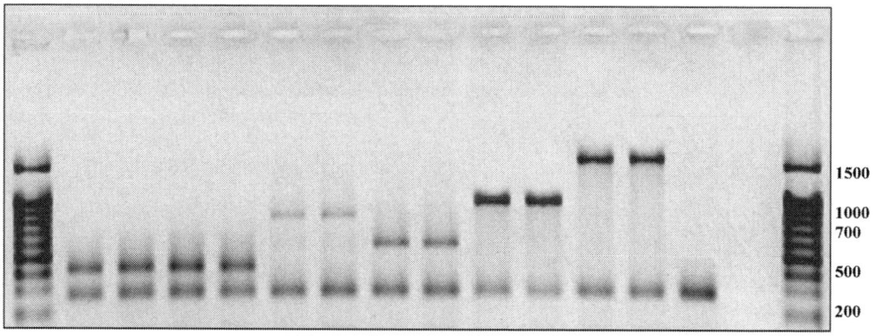

Figure 3. Electrophoretic separation in 2% agarose of the amplicons produced by PCR using four pairs of oligonucleotide primers: (1) CPXV strain GRI-90; (2) CPXV strain Puma-73; (3) CPXV strain Turkmenia; (4) CPXV strain EP-2; (5) MPXV strain CDC# v79-I-005; (6) MPXV strain CDC# v97-I-004 (Central African); (7) MPXV strain CDC# v70-I-187; (8) MPXV strain CDC# v78-I-3945 (Western African); (9) VARV strain Congo-9; (10) VARV strain Ind-3a; (11) VARV strain Butler; (12) VARV strain Brazil-128; (13) ectromelia virus strain MP-1; (14) negative control; and M, DNA marker (lengths in bp are shown to the right).

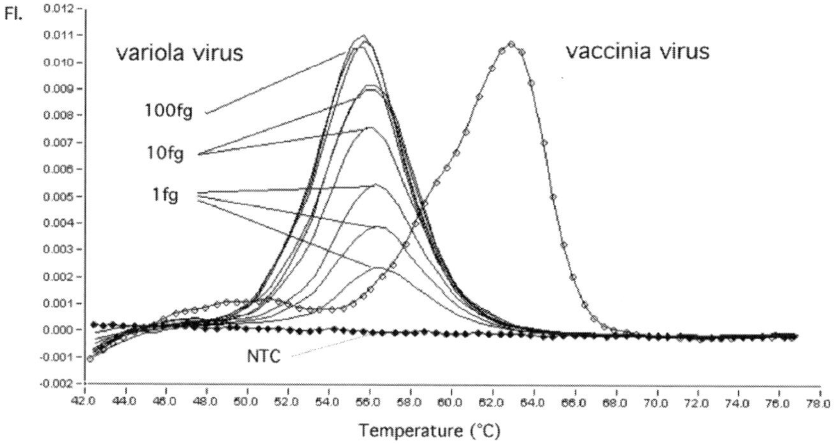

Figure 4. Melting curves generated after LightCycler amplification of 100, 10 and 1 fg DNA prepared from variola virus infected cell culture material (strain Kali Mathu). Each DNA concentration was run in triplicate. The melting curve after amplification of 100 fg of vaccinia virus (strain MVA) is also given [30].

events. Recently, three independent research teams developed the procedures of real-time PCR identification of variola virus and its discrimination from other orthopoxviruses pathogenic to humans [28-30].

Note that screening large orthopoxviral (OPV) strain collections is essential to demonstrate the usefulness, and establish the performance characteristics of assays being developed. In a recent paper [28] the authors state that mismatches in the FRET (fluorescence resonance energy transfer) probes used in their assay enabled discrimination of VARV from other OPVs by DNA melting curve analysis of 204-bp amplicon from the *HA* gene area. Due to new orthopoxvirus sequences in GenBank, the FRET probes display also identity to camelpox and some cowpox virus strains. Analysis of such strains has to prove whether a reliable identification of smallpox virus is still possible.

```
             Праймер TNFR1f
         GCTTCCAGATTATGTGATAGCAAGACTA ->
VARV-IND GCTTCCAGATTATGTGATAGCAAGACTAACACA------CAATGTACACCGTGTGGTTCGGGTACCTTTACATCT 69
VARV-GAR .................................------.................................... 69
MPXV-SL  .............................T...------......................A............. 69
MPXV-ZAI .............................T...------........G.............A............. 69
CPXV-90  ..............C..................AACACA........G.............AC.....C...... 75
CPXV-85  ..............C..................AACACA.............................C...... 75
VACV-COP ..............C..................------........G....................C...... 69
VACV-T-T ..............C..................------........G..C.................C...... 69
CPXV-Utr ..............C..................------.G.........C.................C...... 69
CMLV-CP5 ..............C..................------........G....................C...... 69
CMLV-CP1 ..............C..................------........G..................C.C...... 69
                                  CTAACACAAACACACA               №1
                                  CTAACACA------CGATGTAC         №2
                                  CTAACACA------CAATGTAC         №3

VARV-IND CGCAATAATCATTTACCCGCTTGTCTAAGTTGTAACGGAAGATGCAATAGTAATCAGGTAGAGACGCGATCGTGT 144
VARV-GAR ........................................................................... 144
MPXV-SL  .A..............AG..........................TG............................. 144
MPXV-ZAI .A..............AG..........................TG............................. 144
CPXV-90  .............................................G............................. 150
CPXV-85  .A...........................................G.......................C..... 150
VACV-COP ..........................................----------------------........... 122
VACV-T-T ..........................................----------------------........... 122
CPXV-Utr ..........................................----------------------........... 122
CMLV-CP5 .............................................G............................. 144
CMLV-CP1 .............................................G............................. 144
                     TTACAGGCTTGTCT   №4
                     TTACCCGCTTGTC    №5
                                    CGGAAG------------------------CGCGAT №6
                                       AAGATGTGATAGTAAT                  №7
                                       AAGATGCGATAGTAAT                  №8
                                       AAGATGCAATAGTAAT                  №9

VARV-IND AACACGACTCACAATAGAATCTGTGAATGCTCTCCCGGATATTATTGTCTTCTTAAAGGATCATCCGGATGCAAG 219
VARV-GAR ........................................................................... 219
MPXV-SL  ...................................A.................C......G...TA..G..T.GA 219
MPXV-ZAI ...................................A.................C...........A..G..T.GA 219
CPXV-90  ..........................T...G......................C..................... 225
CPXV-85  .....................................................C..................... 225
VACV-COP .....T...............................................C..................... 197
VACV-T-T .....T...............................................C..................... 197
CPXV-Utr .....................................................C..................... 197
CMLV-CP5 ................................................A....C..................... 219
CMLV-CP1 ................................................A....C..................... 219
                                               GTATTCTCAAAGGA          №10
                                               GTCTTCTCAAAGGA          №11
                                               GTCTTCTTAAAGGA          №12

VARV-IND GCATGTGTTTCCCAAACAAAATGTGGAATAGGATACGGAGTATCCGGA    267
VARV-GAR ................................................    267
MPXV-SL  A.....A....TA.......G...........................    267
MPXV-ZAI A.....A....TA.......G...........................    267
CPXV-90  ......A.........................................    273
CPXV-85  ................................................    273
VACV-COP ................................................    245
VACV-T-T .................C..............................    245
CPXV-Utr ................................................    245
CMLV-CP5 ................................................    267
CMLV-CP1 ................................................    267
               ATTTCTAAAAACAAA      №13
               GTTTCCCAAACAAA       №14
                              <- CCTTATCCTATGCCTCATAGGCCT
                                         Праймер TNFR3r
```

Figure 5. Aligned sequences of a *CrmB* gene fragment of various OPVs. All the sequences are compared with VARV strain IND. Dots indicate identical oligonucleotides; dashes, deletions. Numbers 1 to 14 designate sequences of species-specific probes and their location relative to the aligned sequences. Sequences of the primers TNFR1f and TNFR3r for genus-specific DNA fragment amplification are shown [31].

A new screening assay for real-time LightCycler (Roche Applied Science, Mannheim, Germany) PCR identification of variola virus DNA was developed and compiled in a kit system under GMP conditions with standardized reagents [30]. In search of a sequence re-

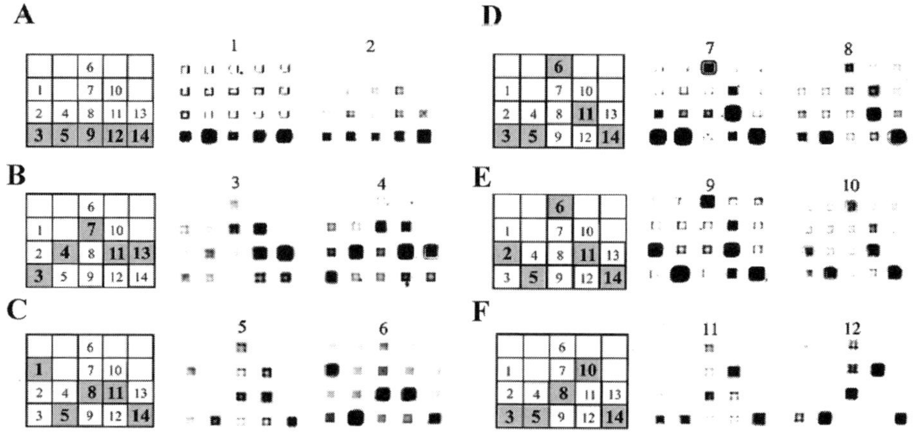

Figure 6. Hybridization patterns obtained on the microchip for five OPV species: A, variola virus (1, strain Garcia-1966 and 2, strain Semat); B, monkeypox virus (3, strain Congo-8 and 4, strain Zaire-96-I-16); C, cowpox virus (5, strain EP-267 and 6, strain GRI-90); D, vaccinia virus (7, strain WR and 8, strain CVI-78); E, rabbitpox virus (9, strain Utrecht and 10, vaccinia virus strain Elstree/Utrecht); and F, camelpox virus (11, strain CP-1 and 12, strain CP-1260/95). Schemes of the anticipated hybridization patterns are shown in the left column; oligonucleotide numbers in the cells correspond to the numbers of oligonucleotides in Fig. 5.

gion unique to variola virus, the nucleotide sequence of the 14 kDa fusion protein gene of 14 variola virus isolates of the Russian WHO smallpox (variola) virus repository was determined and compared to published sequences. PCR primers were designed to detect several species of the genus *Orthopoxvirus*. A single nucleotide mismatch resulting in a unique amino acid substitution in variola virus was used to design a hybridization probe pair with a specific sensor probe that allows reliable differentiation of variola virus from other OPVs via melting curve analysis (Fig. 4). The applicability was demonstrated by successful amplification of 120 strains belonging to orthopoxvirus species variola, monkeypox, cowpox, vaccinia, camelpox, or mousepox virus. The melting temperature (Tm) determined for 46 strains of variola virus (Tm, 55.9-57.8°C) differed significantly (p=0.005) from those obtained for 15 strains of monkeypox virus (Tm, 61.9-62.2°C), 40 strains of cowpox virus (Tm, 61.3-63.7°C), 11 strains of vaccinia virus (Tm, 61.7-62.7°C), 8 strains of mousepox (ectromelia) virus (Tm, 61.9°C), and 8 strains of camelpox virus (Tm, 64.0-65.0°C).

Another highly sensitive and specific assay for the rapid detection of variola virus DNA on both the Smart Cycler and LightCycler platforms was developed [29]. The assay is based on TaqMan chemistry with the orthopoxvirus *HA* gene used as the target sequence. The assay was evaluated in a blinded study with 322 coded samples that included genomic DNA from 48 different isolates of variola virus; 25 different strains of monkeypox, cowpox, vaccinia, rabbitpox, camelpox, ectromelia, gerbilpox, raccoonpox, skunkpox, myxoma, herpes, and varicella-zoster viruses.

Along with the evident advantages of real-time PCR identification, this method yet has certain shortcomings. It allows for analyzing only one relatively short fragment of the virus genome and identifying, as a rule, only one species. Taking into account natural variation of the orthopoxvirus genetic loci analyzed, it is necessary to assay concurrently several individual loci located in different parts of the virus genome. Therefore, a reliable result on identification of a particular orthopoxvirus species requires a parallel or successive real-time PCR identification involving several different genes.

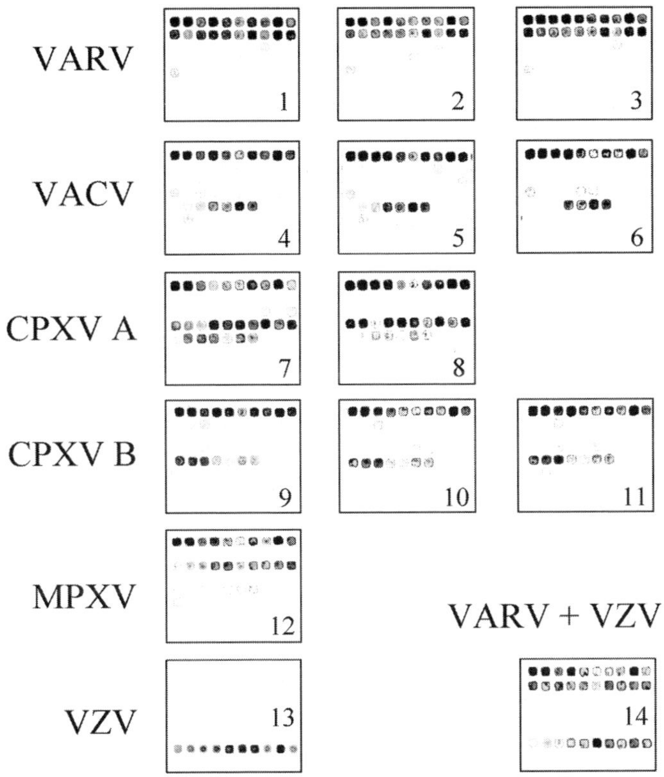

Figure 7. Patterns of detection and discrimination between the OPV species pathogenic for humans and their differentiation from HHV 3 (VZV) by microarray [31]. 1 - VARV strain Congo-2; 2 - VARV strain Ind-3A; 3 - VARV strain Kuw-5; 4 - VACV strain Elstree 3399; 5 - VACV strain Copenhagen; 6 - VACV strain CVI-78; 7 - CPXV-A strain Turk-74; 8 - CPXV-A strain Brighton; 9 - CPXV-B strain EP-1; 10 - CPXV-B strain EP-2; 11 - CPXV-B strain EP-7; 12 - MPXV strain CDC# 77-666; 13 - VZV strain Oka. 14 - Pattern of the mixture hybridization of VARV strain Ind-3A and VZV strain Oka [32].

5. Oligonucleotide Microarray Analysis

Many of the above-mentioned problems that arise during a species-level detection of the viruses can be solved using hybridization of DNA molecules on oligonucleotide microarrays, frequently called microchips. A method for species-specific detection of orthopoxviruses on oligonucleotide microchip was described [31]. The method is based on hybridization of a fluorescently labeled amplified DNA specimen with the oligonucleotide DNA probes immobilized on a three-dimensional polyacrylamide-gel microchip (MAGIChip^{TM}). The probes identify species-specific sites within the viral *CrmB* gene. The microchip contains 14 oligonucleotide probes directed towards 5 species-specific segments of the *CrmB* gene. The probe location relative to the sequence of the gene of the variola virus is shown in Fig. 5.

The microchip contains 5 gel pad columns (Fig. 6), each column representing a separate interrogated segment of the amplified fragment of viral DNA. Within each column, only one species-specific probe can form a perfect duplex with a viral DNA sample, while all other probes form mismatched duplexes. As a result, the hybridization pattern is unique for every tested DNA sample, thus enabling an accurate species assignment.

59 samples of OPV DNA representing six different species were analyzed. Different strains of variola, monkeypox, cowpox, vaccinia, and camelpox viruses were successfully identified using hybridization of amplified DNA to the microchip. No discrepancy between hybridization and conventional identification results was observed.

Another kind of oligonucleotide microarray, created on plain glass slides, was developed for discrimination between orthopoxviruses pathogenic to humans based on viral gene *C23L/B29R* encoding the CC-chemokine binding protein [32].

This microarray-based method detects simultaneously and discriminates four OPV species pathogenic for humans (variola, monkeypox, cowpox, and vaccinia viruses) and distinguishes them from chickenpox virus (varicella-zoster virus or VZV). The microchip contained several unique 13-21 bases long oligonucleotide probes specific to each virus species to ensure redundancy and robustness of the assay. A region approximately 1100 bases long was amplified from samples of viral DNA and fluorescently labeled with Cy5-modified dNTPs, and single-stranded DNA was prepared by strand separation. Hybridization was carried out under plastic coverslips, resulting in a fluorescent pattern that was quantified using a confocal laser scanner (Fig. 7). 49 known and blinded samples of OPV DNA, representing different OPV species, and two VZV strains were tested. The oligonucleotide microarray hybridization technique identified reliably and correctly all samples. This new procedure takes only 3 h, and it can be used for parallel testing of multiple samples.

Acknowledgements

This work was supported in part by ISCT grants #1516p, 1987p, and 2508p.

References

[1] S.S. Marennikova and S.N. Shchelkunov. 1998. Orthopoxviruses Pathogenic for Humans. KMK Scientific Press Ltd, Moscow. 386 pp.

[2] J.G. Breman, D.A. Henderson. 2002. Current concepts: diagnosis and management of smallpox. *N. Eng. J. Med.* **346**, 1300-1308.

[3] R.C. Spencer, N.F. Lightfoot. 2001. Preparedness and response to bioterrorism. *J. Infec.* **43**,104-110.

[4] J.G. Breman. 2000. Monkeypox: an emerging infection for humans? in: W.M. Scheld, W.A. Craig, J.M. Hughes (Eds.), Emerging Infections 4, ASM Press, Washington, pp. 45-67.

[5] V.B. Mukinda, G. Mwema, M. Kilundu, et al. 1997. Re-emergence of human monkeypox in Zaire in 1996. Monkeypox Epidemiologic Working Group. *Lancet.* **349,** 1449-1450.

[6] H. Meyer, C. Schay, H. Mahnel, M. Preffer. 1999. Characterization of orthopoxviruses isolated from man and animals in Germany. *Arch. Virol.* **144**, 491-501.

[7] P. Schupp, M. Pfeffer, H. Meyer, et al. 2001. Cowpox virus in a 12-year-old boy: rapid identification by orthopoxvirus-specific polymerase chain reaction. *Brit. J. Dermat.* **144**, 1-6.

[8] R. Wienecke, H.Wolff, M. Schaller, et al. 2000. Cowpox virus in an 11-year-old girl. *J. Amer. Acad. Dermat.* **42**, 892-894.

[9] C.P. Czerny, A.M. Eis-Hübinger, A. Mayr, et al. 1991. Animal poxviruses transmitted from cat to man: current event with lethal end. *J. Vet. Med.* **138**, 421-431.

[10] CDC. 2003. Update: Multistate outbreak of monkeypox – Illinois, Indiana, Kansas, Missouri, Ohio, and Wisconsin. *MMWR.* **52,** 642-646.

[11] J. J. Esposito, J. C. Knight. 1985. Orthopoxvirus DNA: comparison of restriction profiles and maps. *Virology* **143**:230-251.

[12] J.J. Esposito, J.F. Obijeski, J. H. Nakano. 1978. Orthopoxvirus DNA: strain differentiation by electrophoresis of restriction endonuclease fragmented virion DNA. *Virology* **89,** 53-66.

[13] M. Mackett, L.C. Archard. 1979. Conservation and variation in orthopoxvirus genome structure. *J. Gen. Virol.* **45,** 683-701.

[14] K.D. Reed, J.W. Melski, M.B. Graham, R.L. Regnery, et al. 2004. The detection of monkeypox in humans in the Western Hemisphere. *N. Engl. J. Med.* **350,** 342-350.

[15] A. K. Bej, M. H. Mahbubani, R. M. Atlas. 1991. Amplification of nucleic acids by polymerase chain reaction (PCR) and other methods and their applications. *Crit. Rev. Bioch. Mol. Biol.* **26**, 301-334.
[16] S.N. Shchelkunov, S.M. Resenchuk, A.V. Totmenin, et al. 1993. Comparison of the genetic maps of variola and vaccinia viruses. *FEBS Lett.* **327** 321-324.
[17] R.F. Massung, L.-I. Liu, J. Qi, et al. 1994. Analysis of the complete genome of smallpox variola major virus strain Bangladesh-1975. *Virology.* **201**, 215-240.
[18] S.N. Shchelkunov, A.V. Totmenin, V.N. Loparev, et al. 2000. Alastrim smallpox variola minor virus genome DNA sequences. *Virology* **266,** 361-386.
[19] S.N. Shchelkunov, A.V. Totmenin, I.V. Babkin, et al. 2001. Human monkeypox and smallpox viruses: genomic comparison. *FEBS Lett.* **509**, 66-70.
[20] S.N. Shchelkunov, P.F. Safronov, A.V. Totmenin, et al. 1998. The genomic sequence analysis of the left and right species- specific terminal region of a cowpox virus strain reveals unique sequences and a cluster of intact ORFs for immunomodulatory and host range proteins. *Virology.* **243,** 432-460.
[21] S.J. Goebel, G.P. Johnson, M.E. Perkus, et al. 1990. The complete DNA sequence of vaccinia virus. *Virology.* **179**, 247-266.
[22] G. Antoine, F. Scheiflinger, F. Dorner, F.G. Falkner. 1998. The complete genomic sequence of the modified vaccinia Ankara strain: comparison with other orthopoxviruses. *Virology.* **244**, 365-396.
[23] S.L. Ropp, Q.I. Jin, J.C. Knight, et al. 1995. Polymerase chain reaction strategy for identification and differentiation of smallpox and other ortopoxviruses. *J. Clin. Microbiol.* **33**, 2069-2076.
[24] H. Meyer, S.L. Ropp, J.J. Esposito. 1997. Gene for A-type inclusion body protein is useful for a polymerase chain reaction assay to differentiate orthopoxviruses. *J. Virol. Methods.* **64**, 217-221.
[25] V.N. Loparev, R.F. Massung, J.J. Esposito, H. Meyer. 2001. Detection and differentiation of Old World orthopoxviruses: restriction fragment length polymorphism of the *crmB* gene region. *J. Clin. Microbiol.* **39,** 94-100.
[26] H. Neubauer, U. Reischl, S. Ropp, et al. 1998. Specific detection of monkeypox virus by polymerase chain reaction. *J. Virol Methods.* **74**, 201-207.
[27] E.V. Gavrilova, I.V. Babkin, S. N. Shchelkunov. 2003. A multiplex PCR assay for species-specific rapid identification of orthopoxviruses. *Molekulyarnaya Genetika, Mikrobiologiya i Virusologiya* **1,** 45-52.
[28] M. J. Espy, F. R. Cockerill, F. R. Meyer, et al. 2002. Detection of smallpox virus DNA by LightCycler PCR. *J. Clin. Microbiol.* **40,** 1985-1988.
[29] M.S. Ibrahim, D.A. Kulesh, S.S. Saleh, et al. 2003. Real-time PCR assay to detect smallpox virus. *J. Clin. Microbiol.* **41,** 3835-3839.
[30] V.A. Olson, T. Laue, M.T. Laker, I.V. Babkin, C. Drosten, S.N. Shchelkunov, M. Niedrig, I.K. Damon, H. Meyer. 2004. Real-time PCR system for detection of orthopoxviruses and simultaneous identification of smallpox virus. *J. Clin. Microbiol.* (in press).
[31] S. Lapa, M. Mikheev, S. Shchelkunov, V. Mikhailovich, A. Sobolev, V. Blinov, L. Babkin, A. Guskov, E. Sokunova, A. Zasedatelev, L. Sandakhchiev, A. Mirzabekov. 2002. Species-level identification of orthopoxviruses with an oligonucleotide microchip. *J. Clin. Microbiol.* **40,** 753-757.
[32] M. Laassri, V. Chizhikov, M. Mikheev, S. Shchelkunov, K. Chumakov. 2003. Detection and discrimination of orthopoxviruses using microarrays of immobilized oligonucleotides. *J. Virol. Methods.* **112,** 67-78.

Radiation Inactivation of Bioterrorism Agents
L.G. Gazsó and C.C. Ponta (Eds.)
IOS Press, 2005

Inactivation of Bio-Terrorism Agents in Military and Domestic Applications

Timothy G. HENRY
President, BioSecurity, Ion Beam Applications Inc.,
95 Mason Drive, Princeton, New Jersey, USA

Abstract. The Anthrax attacks in the Fall of 2001 resulted in a heightened awareness of the role radiation plays in the inactivation of BioTerrorism Agents. After a review of thousands of proposals to decontaminate mail, only X-ray and Electron Beam solutions were implemented by the United States Postal Service. A brief overview of the current process used to sanitize the US mail is presented.

Often over time, radiation solutions for industrial, scientific, and government applications are displaced by less sophisticated alternatives. A comparison of the radiation solution to other technologies is presented to suggest why, in the case of BioTerrorism, the radiation solution is here to stay. Military, homeland defense, mail, and commercial applications for radiation inactivation of Bio-Terrorism agents are growing. One of those applications is also presented here.

Introduction

For the past few decades, the best and brightest individuals have been attracted to the radiation field. As a result, when a new problem arises a radiation solution is one of the first to be proposed. Sometimes the radiation solution persists, e.g., for medical device sterilization. In other instances, less expensive solutions eventually become the preferred solution, e.g., sewage sludge remediation. Thus, before committing a significant scientific effort into developing radiation sanitization, it is important to compare practical advantages of all the technologies for inactivating Bio-Terrorism agents such as *Bacillus anthracis* spores and Ricin.

The first application that is described is the use of X-rays and Electron Beams to sanitize mail. While the process details cannot be shared publicly, a general description of how mail flows through a facility is instructive. The author will use the mail application to compare the available technologies.

Radiation sanitization is not a complete solution for all biological and chemical threats nor is it without collateral damage to the product being sanitized. However, by understanding the mechanism of inactivation, it is hoped that radiation sanitization can be extended to Bio-Terrorism agents other than *Bacillus anthracis* spores (Anthrax). This author believes radiation will continue to be the most effective mechanism for depositing the energy required to accomplish inactivation, but that other synergistic technologies will be necessary to extend to the process to other Bio-Terrorism agents.

The use of radiation to protect the mail has led to numerous other possible applications. One of these applications, corpse irradiation for the US military, has led to the development of a mobile platform, which could have a far reaching application for deploying sanitization capability to contaminated sites regardless of the location or the nature of contaminated material.

1. Mail Irradiation

Mail irradiation is the best-known example of the use of radiation to inactivate Bio-Terrorism agents. It is not all that dissimilar from the radiation sterilization of medical devices, yet mail sanitization differs from medical device sterilization in a number of fundamental ways.

- Medical devices are generally clean and relatively low in bioburden.
- The density and composition of the product to be sterilized is known.
- The biological threat and its response to radiation are known.
- There are generally accepted guidelines for processing.
- Workers and facilities are not put at risk from the products being sterilized.

Despite the differences, there are many lessons that have been learned over the decades of medical device sterilization that prove useful in the mail sanitization application.

1.1. History

In the fall of 2001, a small number of letters were mailed to public and private citizens that contained *Bacillus anthracis* spores. Per a US Government Accounting Office report [1], "as of December 6, 2001 the known facts included:

- 4 known letters containing anthrax
- 11 confirmed cases of inhalation anthrax; 5 deaths
- 23 USPS facilities found to be contaminated
- Approximately 1.8 million pieces of mail requiring decontamination
- Anthrax Spore diameter is 1 micron"

In response to the Anthrax attack, the United States Postal Service (USPS) contracted with IBA to develop a process for the decontamination of quarantined mail. Subsequently, a mail sanitization process was developed to protect mail destined for specific facilities in Washington, DC, from future Anthrax attacks. The involvement of IBA in this process came as the result of its role as a global innovator in the design and development of particle accelerators and its position as the world's largest contract medical device sterilization provider. [2] Its broad base in other sterilization technologies complements the IBA radiation expertise, and these technologies were evaluated by IBA before making a technology recommendation.

1.2. Technology Comparison

The first consideration for implementing a sanitizing technology is a practical one. The penetrating ability of the technology is compared to the requirement. From Figure 1, it can be seen the UV light might be a candidate for surface treatment of envelopes but not for the inside of envelopes. Similarly X-rays would be very efficient if mail is processed in bulk, but the penetration capability of X-rays would be wasted if the surfaces of individual envelopes were presented one-by-one in an X-ray process.

Also evident from Figure 1 is the implication that as the volume of mail increases, certain technologies are preferred. For example, traditional linac X-ray units could process a single

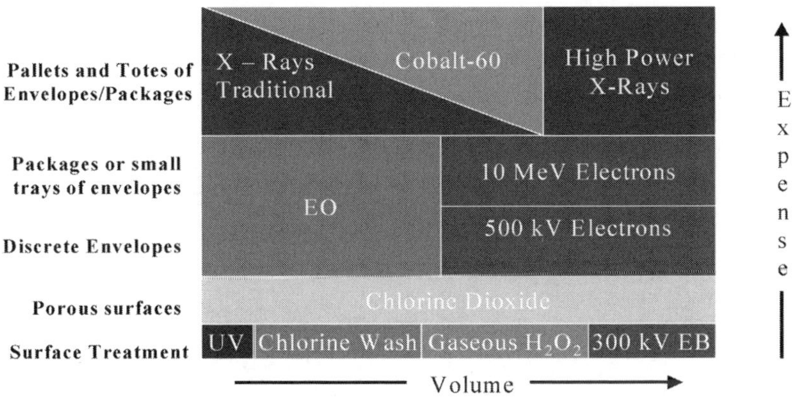

Figure 1. Technology Comparisons for Mail.

small package of mail containing envelopes in an hour or two. For larger volumes of mail, high-power units like the one located in IBA's Bridgeport, NJ, facility are required.

On a practical basis, only X-ray, Electron Beam, and Gamma facilities are able to process the high volume of mail required for the USPS volumes. However, there are other limitations, which favor X-ray as the ideal solution.

The unknown nature of mail contents make it necessary to plan for a wide range of materials which may vary widely in density and composition. Since anything can be turned into mail by placing a stamp on it, these unknowns can create additional problems for non-X-ray solutions.

- Possibility of bombs in the mail compounded by their explosive interaction with the sanitization system – radioactive gamma ray emitters, ethylene oxide, and chlorine dioxide
- Destructive nature of heat, moisture, or chemicals on the functionality of the mail -heat, steam, electron beam, chlorine dioxide, chlorine wash
- Barriers that impede or prevent the sanitizing agent from penetrating throughout - all technologies except heat, Gamma rays, and X-rays
- Existence of a protective characteristic of *Bacillus anthracis* spores that results in an inactivation bio-asymptote, i.e., the existence of the sanitization plateau for at least one of the non-radiation sanitization techniques

1.3. Bridgeport, NJ, Mail Sanitization Facility

Together with the USPS, the National Institute for Standards and Technology (NIST), the White House Office of Science and Technology Policy (OSTP), and the Armed Forces Radiobiology Research Institute (AFRRI), IBA developed protocols for the use of its Bridgeport, NJ, Electron Beam and X-ray facility (Figure 2) to sanitize thousands of pounds of mail on a daily basis. This facility is now used to protect the White House, Congress, and various federal agencies in or near Washington, DC.

Figure 2 shows a plant that is actually two plants in one: an X-ray facility on the right side and Electron Beam facility on the left side. Built originally for processing food and plastics, the facility has been modified to sanitize single-envelope-high boxes in the electron mode, and mail in large stainless steel totes in the X-ray mode.

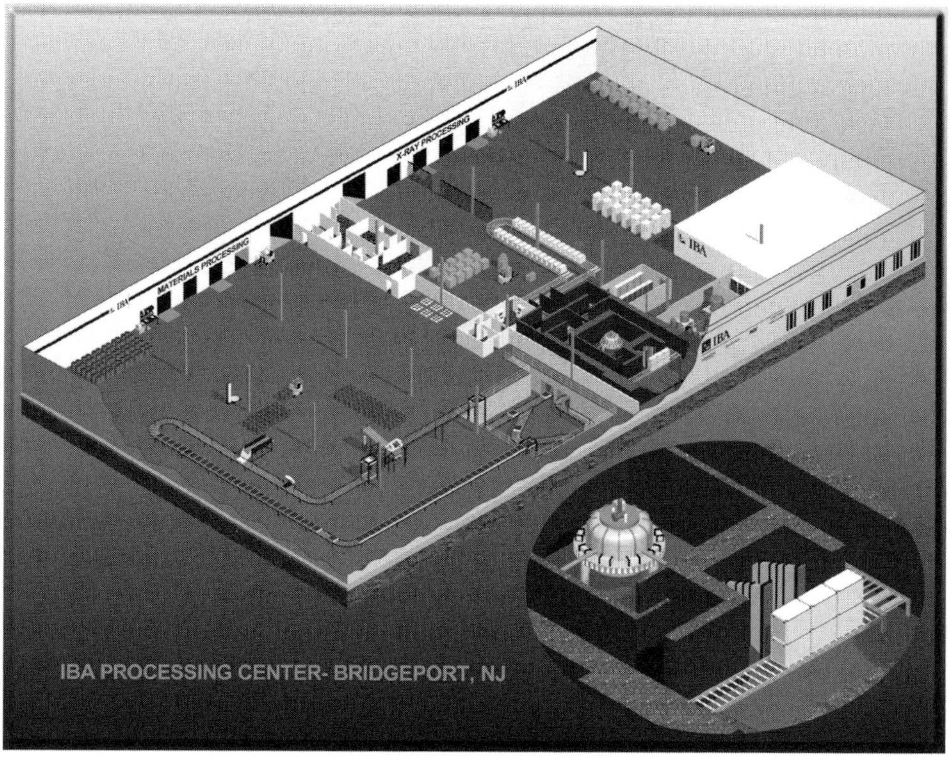

Figure 2. Mail Processing Facility with X-ray cut-away.

Mail enters the building through a specially designed isolation room, which itself is monitored for Bio-Terrorism agents. Depending on the physical form of the mail, it is sent to either the 10 MeV Electron Beam chamber or to the X-ray chamber, which has both a 5 and a 7 MeV X-ray capability. In the cutaway of Figure 2 it is possible to see the stacking of metal totes used in the X-ray portion of the facility. Upon exiting the sanitization chamber, the mail containers are returned to the USPS. At no time is it necessary for IBA personnel to handle or inspect the mail.

1.4. Future Mail Process Enhancements

The science to deposit energy at the molecular level to inactivate Bio-Terrorism agents already exists. The energy required can be provided to the molecules via chemicals, radiant heat, or radiation, etc. The beauty of radiation is that energy is transferred directly to the molecules, regardless of packaging barriers. It is difficult to imagine that there is a more efficient way than irradiation to transfer this energy. A challenge for the future is find ways to impart just enough energy to inactivate the biological or chemical threats without causing collateral damage to the items being sanitized.

In a fundamental view of radiation sanitization, the inactivation of Bio-Terrorism agents is accomplished by providing enough energy to a substrate to allow it to destroy itself. As molecules attempt to dissipate energy, one pathway that is open is the making and breaking of chemical bonds, e.g., in a *Bacillus anthracis* DNA molecule. In addition, molecular torsion or rotation of bonds can reshape a protein destroying a catalytic effect, e.g., in the enzymatic area

Figure 3. Transportable Biowarfare Agent Defeat System (TBADS™).

of a Ricin molecule. A fruitful area of research for the future may be the blending of technologies to ensure that less energy is dissipated as heat and more energy is diverted to one of energy pathways that lead to Bio-Terrorism agent inactivation. To extend the process used for Anthrax to Ricin, IBA is currently evaluating a blending of sanitization technologies such that energy imparted by radiation is trapped in the molecules until inactivation occurs.[3]

2. Military Applications

2.1. Corpse Irradiation

The successful implementation of X-rays for sanitizing the mail led to evaluations of the technology for non-mail applications. Fallen soldiers on the battlefields who are contaminated with Bio-Terrorism agents cannot be safely transported home for burial until the biological agents are inactivated. The option of cremating soldiers' remains is unacceptable to the US military and to many cultures around the world. To address this need, IBA worked with US Department of Defense to develop a mobile battlefield X-ray sanitizer (Figure 3).[4]

Monte-Carlo computer simulations of the mobile sanitizer were validated by running trials with test dummies in battle gear through the X-ray fields at the Bridgeport facility. The empirical data was then used to calibrate the Monte-Carlo modeling. Excellent agreement between the computer modeling and measured dosimetry was achieved.

3. Conclusion

Other applications of radiation inactivation of Bio-Terrorism agents that are already being evaluated include decontamination of shipping containers at ports, contaminated water, contaminated soil, and passenger luggage. With the success experienced to date with mail sanitization, the development of a mobile unit, and the supporting inactivation studies, it is easy to imagine that radiation will continue to be a valuable tool in the defense against Bio-Terrorism agents.

References

[1] DIFFUSE SECURITY THREATS, Technologies for Mail Sanitization Exist, but Challenges Remain, United States Government Accounting Office report GAO-02-365, April 2002, p. 9, Washington, DC.
[2] IBA 2002 Annual Report, inside front cover, available from IBA via Website: www.iba-worldwide.com.
[3] IBA BioSecurity Addresses Ricin Scare, IBA press release, February 9, 2004, Louvain-La-Neuve, Belgium.
[4] Killing Anthrax, Government Security News, October, 2003, Vol. 1 Issue 2, p. 18, New York, NY.

Inactivation of Biological Warfare Agent Simulants by Ionizing Radiation

Thomas B. ELLIOTT, PhD, Gregory B. KNUDSON, PhD,
Michael O. SHOEMAKER, PhD and G. David LEDNEY, PhD
Armed Forces Radiobiology Research Institute,
8901 Wisconsin Avenue, Bethesda, MD 20889-5603, USA
TEL: (301) 295-0898, FAX: (301) 295-6503

Abstract. The Armed Forces Radiobiology Research Institute (AFRRI) in the U. S. Department of Defense conducts biomedical research on the effects of ionizing radiation. It has the largest radiobiology program in the United States and is a national resource in the response to nuclear and radiation accidents. Bacterial spores are potential biological weapons because they can be prepared and distributed by aerosol, they endure harsh environmental conditions, and they are infectious. Decontamination procedures for large concentrations of spores must be effective and practical. We determined the dose response of bacterial spores to three qualities, or types, of ionizing radiation.

Inactivation of dry and hydrated bacterial spores with gamma radiation has been more thoroughly studied than spore inactivation with neutron radiation. Decimal-reduction curves were produced at AFRRI for *Bacillus atrophaeus* (*B. subtilis* var. *niger, B. globigii*, "BG"), *B. pumilus*, *B. thuringiensis*, and *B. anthracis* Sterne spores, both wet and dry, using doses of 0.3 to 7.2 kGy neutrons delivered at a dose rate of 44 to 49 Gy/min (D_n/D_T = 0.95) in the AFRRI training, research, isotope-producing General Atomic (TRIGA) Mark-F nuclear reactor, and doses of 0.6 to 24.0 kGy gamma rays delivered at dose rates of 112 to 120 Gy/min in the AFRRI cobalt-60 (^{60}Co) gamma-photon irradiation facility. Decimal-reduction curves were constructed by plotting the spore survival fraction in terms of colony-forming units *vs*. radiation dose. All four species showed greater sensitivity to neutron radiation than to gamma radiation, regardless of the state of hydration. Dry spores of all four species were more sensitive to gamma radiation than were hydrated spores. In contrast, the state of hydration, whether dry or hydrated, of spores of *B. subtilis* and *B. pumilus*, which were embedded in filter paper strips, did not affect their sensitivity to neutron radiation. Wet *B. thuringiensis* spores were only slightly more sensitive to neutrons than were *B. thuringiensis* spores in dry powdered form. Furthermore, the species most resistant to neutron and gamma radiations was a concentrated *B. anthracis* Sterne spore suspension. When the starting spore concentration, the *Bacillus* species used, the radiation quality (neutron or gamma), and the state of hydration are known, the radiation decimal-reduction curves generated in these studies can be used to predict bacterial spore survival.

Electron-beam radiation (e-beam) has been used to inactivate microorganisms in spices, fresh food, medical components, and hazardous waste. AFRRI assessed the efficacy of using an e-beam for decontamination of bulk biological agents and of byproducts of the decontamination procedures such as wipes and aqueous runoff. Biological agent surrogates were tested under controlled conditions to determine the effectiveness of e-beam for decontamination. Using the AFRRI linear accelerator (LINAC) to deliver doses of 2 to 20 kGy at a dose rate of 1 kGy/min, radiation decimal-reduction curves were constructed for *Bacillus atrophaeus* spores in a dry powder and *B. anthracis* Sterne spores in a slurry. Doses of 0.25 to 1.0 kGy were delivered to vegetative Gram-negative bacterial cells of *Serratia marcescens*. The LINAC produced 13-MeV electrons at 30 pulses/sec with a 4-μsec pulse width generated through a water scatterer. Spore samples were irradiated in an array of three screw-capped polystyrene tubes. Dosimetry was performed at the beginning of each experimental run with LiF:Ti,Mg thermoluminescent dosimeters (TLDs), product type TLD-100 (Bicron®). TLDs were processed with the Bicron®/Harshaw Model 5500 Automatic TLD Reader. The inactivation data for dry *B. atrophaeus* spores, *B. anthracis* Sterne spores, and *S. marcescens* were fitted to a mathematical formula. The e-beam decimal-reduction curves for the bacterial spores were similar to our previously generated gamma-photon radiation curves. The vegetative bacterial cells of *S. marcescens* were more susceptible to high-speed electrons than were the bacterial spores. These experimental findings support the concept of using a truck-mounted transportable LINAC in the field for decontaminating bulk materials that are contaminated with pathogenic bacteria.

1. Introduction

A brief overview of the mission, capabilities, and radiation resources of the Armed Forces Radiobiology Research Institute (AFRRI) is presented followed by a summary of data, which are derived from several studies, to estimate the decimal reduction (D_{10}) values for inactivation of bacterial endospores of *Bacillus* species by three qualities of ionizing radiation, gamma photons, fast neutrons, and high-energy electrons. The D_{10} value for inactivation of vegetative, Gram-negative cells of *Serratia marcescens* by high-energy electrons are also presented for comparison.

1.1. Mission, Capabilities, and Resources of AFRRI

The principal mission of the AFRRI is to support the U.S. Department of Defense in medical, nuclear, and radiological readiness. The AFRRI conducts radiobiology research and develops medical countermeasures for the U.S. Department of Defense from three perspectives – prevention, assessment, and treatment. AFRRI also trains medical personnel on the medical effects of ionizing radiation; deploys a Medical Radiobiology Advisory Team to nuclear incidents; and advises the Joint Chiefs of Staff, particularly the J4 Medical Officer, the Deputy Assistant Secretary of Defense for Nuclear Matters, the commanders-in-chief, and others on radiological matters.

The Radiation Sources Department of AFRRI operates four radiation sources. The training, research, isotope-producing – General Atomic (TRIGA) Mark-F reactor operates at pulses of up to 2,500 megawatts (MW) and at a steady rate of 1 MW. The cobalt-60 (^{60}Co) irradiation facility provides researchers with movable sources, large uniform gamma-ray fields, and a wide array of exposure configurations. The linear accelerator (LINAC) generates high-energy electrons up to 54 MeV and x-rays. These three sources were used to assess inactivation of BW agents. AFRRI also has a ^{60}Co low-level irradiation facility, which delivers chronic radiation doses to biological samples, to study early and late effects.

The AFRRI Veterinary Sciences Department administers an internationally accredited animal care and use facility and program (AAALAC-I). The 32,000-ft^2 animal facility was designed to support radiation and surgical studies for AFRRI, the U.S. Food & Drug Administration, the U.S. Navy, Walter Reed Army Medical Center, and the Uniformed Services University of the Health Sciences.

The four core research thrusts in AFRRI are Radiation Infection Treatment, which evaluates and develops innovative therapeutic strategies for infections after irradiation; Radiation Casualty Management, which evaluates and develops radioprotective treatments; Biological Dosimetry, which develops biological procedures for assessing radiation doses received by persons, who are involved in nuclear incidents; and Health Effects of Embedded Heavy Metals, which assesses the toxicity associated with depleted uranium and other metals of military relevance.

1.2. Experimental Studies

In studies of the inactivation of bacterial spores by ionizing radiation several variables were examined including radiation source (*e.g.*, an underground nuclear test *cf.* reactor); radiation quality (*e.g.*, neutrons *cf.* ^{60}Co gamma photons); radiation dose and dose rate (*e.g.*, pulsed *cf.* protracted delivery); biological warfare agent simulants (*e.g.*, Gram-positive, spore-forming *Bacillus* species, Gram-negative bacteria, and viruses); and state of hydration (*i.e.*, dry spores *cf.* wet spores).

Simulants used in these studies for virulent *Bacillus anthracis* endospores included *B. anthracis* Sterne, *B. atrophaeus* (*B. subtilis* var. *niger*), *B. pumilus,* and *B. thuringiensis.* The simulant used for virulent *Yersinia pestis*, which causes plague, and *Francisella tularensis*, which causes tularemia, was the vegetative Gram-negative species, *Serratia marcescens*, which produces a distinctive red pigment and was used during the 1950's as a presumably non-virulent environmental test agent [3].

Bacillus atrophaeus ATCC 9372 has been used extensively for environmental testing including open-air field studies. This species was previously classified taxonomically as *B. subtilis* var. *niger*, which was originally described as *B. globigii* [5, 7]. We used a dry, free-flowing preparation of spores at a high concentration of more than 1×10^{11} CFU/g.

Bacillus anthracis Sterne forms spores, which are identical to those of virulent strains, and has the pXO1 plasmid, which encodes the ability to produce the protective antigen, edema factor, and lethal factor; however, this strain does not possess the pXO2 plasmid, which encodes the ability to produce the D-glutamic-acid capsule that resists phagocytosis by macrophages. Live Sterne strain spore vaccines are licensed in the United States for the protection of domestic livestock against anthrax. The Sterne strain is a natural isolate of *B. anthracis* resulting from the spontaneous loss of the pXO2 plasmid. The strain was isolated by Max Sterne and developed into a live attenuated anthrax vaccine in 1939 at the Veterinary Research Institute in South Africa.

1.3. Objective

The objectives of the studies reported here were (1) to evaluate the relative efficacy of three qualities of ionizing radiation – fast neutrons, gamma photons, and high-energy electrons – as a means to inactivate bacterial endospores and vegetative cells for the purpose of using irradiation as a decontamination procedure; and (2) to evaluate several species of *Bacillus* spores and *S. marcescens* for their suitability as surrogates for potential BW agents.

2. Materials and Methods

2.1. Bacterial Spores

Bacillus atrophaeus ATCC 9372 and *B. pumilus* ATCC 27142 spore-impregnated cellulose fiber strips were obtained from Raven Biological Laboratories (Omaha, Nebraska, USA). The strips were made of Schleicher and Schuell #470 filter paper and measured 6 mm × 38 mm × 1 mm. They contained a bioburden of 1×10^8 *B. atrophaeus* spores or 1.3×10^8 *B. pumilus* spores. Before and after irradiation, the packages, which contained the spore strips, were stored in a dehumidified environmental chamber with Drierite (W. A. Hammond Drierite Co., Xenia, Ohio, USA) in a refrigerator at 4°C.

Bacillus anthracis Sterne spores were produced at AFRRI in 1-liter batch fermentations using Schaeffer's sporulation medium inoculated with anthrax spore vaccine (Colorado Serum Co., Denver, Colorado). A wet suspension of centrifuged and washed spores in sterile water, or slurry, contained approximately 3×10^9 CFU/ml. To prepare spore strips, a 0.1-ml volume of the slurry was inoculated to a sterile blank filter-paper strip and allowed to dry in a sterile Petri dish in a dessicator.

Bacillus thuringiensis subsp. *kurstaki* spores were produced by Abbott Laboratories, North Chicago, Illinois, as technical grade DiPel® powder. DiPel® is a commercial insecticide, which consists of *B. thuringiensis* spores and a proteinaceous crystalline toxin, delta-endotoxin. To prepare a wet suspension, or slurry, sterile water was added to an amount of dry powdered spores.

2.2. Preparation of Spores for Irradiation

2.2.1. Dry Spores

Each spore strip was aseptically removed from its envelope and 8 to 10 dry strips were packed into round-bottomed 12 × 75-mm sterile polystyrene test tubes. Before packing the spore strips, sterile cotton-dressing gauze sponges were forced in the bottom 30 mm of each tube. The gauze sponges have radiation-absorption and -scattering characteristics that are equivalent to those of filter paper and they displaced air in the bottom of the tubes. By placing the spore strips at the top of the cotton plug, the position of the spore strips within the tubes was reproducible for irradiation. A 2.6-gram amount of dry *B. thuringiensis* spores (DiPel) was placed in tubes.

2.2.2. Wet Spores

One spore strip was placed in 4 ml sterile water in each 12×75-mm sterile polystyrene test tube. A 4-ml volume of concentrated suspensions of *B. anthracis* Sterne or *B. thuringiensis* was placed in tubes.

2.3. Quantitation of Spore Survival

The technique for quantifying the total number of viable spores was adapted from the U. S. Pharmacopoeia and National Formulary XXII Monograph 1035, 1990 (pp.170-175). Spore strips were homogenized in sterile blender cups in sterile water. Suspensions were heated for 15 min at 65-70ºC. to enhance germination of spores and to kill vegetative bacteria. The suspensions were then cooled in an ice bath at 0-4ºC. and were diluted ten-fold in sterile water. Pour-plates were prepared in duplicate 100-mm Petri dishes with 1 ml of each dilution in casein-soybean digest agar (Trypticase Soy Agar 11043, Becton Dickinson, Cockeysville, Maryland, USA). Following high doses of radiation, the entire suspension of homogenized spore strips was inoculated in 5-ml aliquots into 150-mm pour plates. Solidified agar plates were incubated inverted for 48 hr at 30-35ºC. Colonies were counted at 48 hr.

2.4. Irradiation

2.4.1. Neutron Irradiation

Bacterial spore preparations in sterile, 12×75-mm, polystyrene test tubes were loaded into a Lucite holder (measuring 23 cm × 7.5 cm × 7.5 cm), which held up to 12 test tubes in a single row. The tubes were suspended between two 3.25-mm-thick Lucite restrainer walls. Empty tubes were placed in positions at each end of the array to ensure similarity of scatter to all test tubes. The exposure array was placed in a lucite harness, which was passed through an aluminum extractor tube into a bismuth cave designed to reduce the γ-photon component of the total dose of radiation given to the spores in an exposure room of the AFRRI TRIGA reactor. Detailed characteristics and dosimetry have been described [4, 6]. At the position of irradiation, the measured ratio of the neutron dose to the total dose was 0.95±0.02 delivered at a steady-state reactor power of 500 kW and a dose rate of 46.2 Gy/min (range 44-49 kGy/min). Doses delivered to samples were between 0.3 and 7.2 kGy. The average energy of fast neutrons was 0.71 MeV. Temperature in the exposure room was 25±2°C. The uniformity of the radiation field was within ±6% of the dose at the center of the array.

2.4.2. Gamma Irradiation

Bacterial preparations in sterile, 12 × 75-mm, polystyrene test tubes were loaded into a lucite holder (measuring 7.5 cm × 15 cm with 6-mm walls), which held 10 tubes in a single row. Empty tubes were placed in positions at each end of the array to ensure identical conditions for all spore strips in the tubes. The tubes were exposed bilaterally to γ-photon radiation in the AFRRI ^{60}Co irradiation facility [2]. The dose rate was measured before irradiation of bacteria under conditions that were identical to those for irradiating the bacteria by placing a 0.5-cm^3 tissue-equivalent ionization chamber in a test tube at a known distance from the ^{60}Co source. The ratio of the exposure rate measured in test tubes to the dose rate in free air for this array was 0.95. Exposure time was adjusted so that the bacteria received the desired dose at a nominal dose rate of 112-120 Gy/min. Doses delivered to samples were between 0.6 and 24.0 kGy. Variation of dose within the exposure field was ±2%. The average energy of gamma photons was 1.2 MeV. Temperature in the exposure room was 21±1°C. The techniques used for these measurements were performed in accordance with the protocol of the American Association of Physicists in Medicine for the determination of absorbed dose from high-energy photon and electron beams [1].

2.4.3. High-energy Electron-beam Irradiation

The AFRRI linear accelerator produced 13-MeV electrons at 30 pulses/sec with a 4-μsec pulse width generated through a water scatterer. Dosimetry was performed at the beginning of each experimental exposure by using LiF:Ti,Mg thermoluminescent dosimeters (TLDs), product type TLD-100 (Bicron®). TLDs were processed with the Bicron®/Harshaw Model 5500 Automatic TLD Reader. The dose rate was 1 kGy/min. Doses delivered to samples were between 0.25 and 20 kGy.

3. Results

3.1. Objective: Inactivation of BW Agent Simulants with Fast Neutrons, ^{60}Co Gamma Photons, and High-speed Electrons

Currently available decimal reduction (D_{10}) values are presented in Table 1. Neutron-irradiated dry spore strips of *B. anthracis* Sterne were more resistant than *B. pumilus*, *B. atrophaeus*, or powdered spores of *B. thuringiensis*. Similarly, a neutron-irradiated wet slurry of *B. anthracis* Sterne spores was more resistant than wet spore strips of *B. pumilus* and *B. atrophaeus*, or a wet slurry of *B. thuringiensis* spores. Wet spores of *B. anthracis* Sterne were more resistant than dry spores.

On the other hand, gamma-photon-irradiated dry spore strips of *B. pumilus* were more resistant than those of dry powder of *B. thuringiensis* spores or dry spore strips of *B. anthracis* Sterne and B. *atrophaeus*. However, gamma-photon-irradiated wet spore strips of *B. anthracis* Sterne and *B. pumilus* were more resistant than wet spore strips of *B. atrophaeus* and a wet slurry of *B. thuringiensis* spores. A wet slurry of *B. anthracis* Sterne spores was more resistant than the other four spore preparations.

Table 1. Decimal reduction (D_{10}) values for dry bacterial spores irradiated with fast neutrons, ^{60}Co gamma photons, or high-speed electrons. (Armed Forces Radiobiology Research Institute).

Radiation Quality	Microorganism	D_{10} (kGy)[d]	
		Dry	Wet
Gamma	*B. atrophaeus*[a]	1.1, 5.7[e]	2.1
(112-120 Gy/min)	*B. atrophaeus*[b]	2.0	–
	B. anthracis Sterne[a]	1.1, 5.7[e]	2.7
	B. anthracis Sterne[c]	–	4.3
	B. pumilus[a]	1.4	2.8
	B. thuringiensis[b]	1.2	2.0
Neutron	*B. atrophaeus*[a]	0.5	0.5
(46.2 Gy/min, D_n/D_γ = 0.95)	*B. anthracis* Sterne[a]	0.9	1.1
	B. pumilus[a]	0.7	0.7
	B. thuringiensis[b]	0.4	0.3
Electrons	*B. atrophaeus*[b]	1.3, 3.3[e]	–
(1.0 kGy/min)	*B. anthracis* Sterne[c]	–	3.3
	Serratia marcescens	–	0.04, 0.2[e]

[a]filter paper strip, 10^8 CFU/strip
[b]0.5 g powder/sample
[c]wet slurry of centrifuged and washed spores
[d]D_{10} value: Radiation dose required to kill 90% of the population (*i.e.*, 1-logarithm reduction)
[e]two portions of a hyperbolic curve (The first value is calculated from the declining portion of the curve with a steep slope and the second value is calculated from the portion of the curve at the higher doses of radiation with a shallow slope, which indicates that a small proportion of the microorganism is resistant.)

Decimal reduction (D_{10}) values for dry spores of *B. atrophaeus*, *B. anthracis* Sterne slurry, and *S. marcescens* Gram-negative vegetative cells irradiated with high-energy electrons are presented in Table 1. The log-linear survival curves generated for the bacterial spores, irradiated in the AFRRI LINAC, had slopes similar to those previously generated using ^{60}Co gamma-photon irradiation. The cells of *S. marcescens* were considerably more susceptible to inactivation by high-energy electrons than were the bacterial spores.

4. Conclusions

- Gamma photons, neutrons, and high-speed electrons effectively inactivated bacterial spores and vegetative bacterial cells.
- Concentrated spores of *B. anthracis* Sterne in aqueous suspension (slurry) were the most resistant of all four species to both gamma-photon and neutron radiation.

- Spores of *B. atrophaeus*, *B. anthracis*, *B. pumilus*, and *B. thuringiensis* were more susceptible to neutron radiation than to gamma-photon or high-speed-electron radiation.
- Dry spores of *B. atrophaeus*, *B. anthracis*, *B. pumilus*, and *B. thuringiensis* were more susceptible than hydrated spores to gamma-photon radiation.

5. Summary and Recommendations

- Ionizing radiation is a proven method of decontamination and sterilization and it has the unique advantage of penetration.
- AFRRI possesses several sources of ionizing radiation that can be used to study inactivation of biological threat agents.
- Field studies should be conducted to validate the use of a fixed-site electron-beam system to establish field decontamination parameters, followed by studies to validate a transportable e-beam decontamination system.
- These tests should include boxes of bulk materials, which contain sealed packets of surrogate *Bacillus* sp. spores.
- Definitive follow-on studies are recommended to assess the feasibility of:
  (1) e-beam decontamination of heat-sensitive equipment and materials and
  (2) sterilizing BW-agent-contaminated human remains for repatriation by using ionizing radiation.
- Preliminary tests in these two areas have been completed at AFRRI.

Acknowledgements

These studies were supported by RWU00129 and RWUNBC0101, Armed Forces Radiobiology Research Institute, Bethesda, Maryland. The views, opinions, and findings contained in this report are those of the authors and do not reflect official policy or positions of the Department of the Army, the Department of Defense, or the U.S. Government. The following persons contributed to these studies: Gregory B. Knudson, Ph.D., G. David Ledney, Ph.D., Rita A. Harding, M.S., Eric E. Kearsley, Ph.D., Thomas B. Elliott, Ph.D., Michael O. Shoemaker, Ph.D., J. H. Thakar, Ph.D., William E. Jackson, III, the radiation sources staff, and the graphic arts staff.

References

[1] American Association of Physicists in Medicine. 1983. Task Group 21, Radiation Therapy Committee, A protocol for the determination of absorbed dose from high-energy photon and electron beams. Medical Physics **10**:741-771.

[2] Carter, R. E., and D. M. Verrelli. 1973. AFRRI cobalt whole-body irradiator. Technical Note 73-3. Armed Forces Radiobiology Research Institute.

[3] Cole, L. A. 1988. Clouds of Secrecy. The Army's Germ Warfare Tests over Populated Areas. Rowman and Littlefield, Totowa, New Jersey.

[4] Elliott, T. B., G. D. Ledney, R. A. Harding, P. L. Henderson, H. M. Gerstenberg, J. R. Rotruck, M. H. Verdolin, C. M. Stille, and A. G. Krieger. 1995. Mixed-field neutrons and γ photons induce different changes in ileal bacteria and correlated sepsis in mice. International Journal of Radiation Biology **68**:311-320.

[5] Fritze, D., and R. Pukall. 2001. Reclassification of bioindicator strains *Bacillus subtilis* DSM 675 and *Bacillus subtilis* DSM 2277 as *Bacillus atrophaeus*. International Journal of Systematic and Evolutionary Microbiology **51**:35-37.

[6] Ledney, G. D., G. B. Knudson, R. A. Harding, R. C. Bhatt, E. E. Kearsley, and J. A. Zmuda. 1996. Neutron and γ-ray radiation killing of *Bacillus* species spores: dosimetry, quantitation, and validation techniques. Technical Report 96-1. Armed Forces Radiobiology Research Institute.
[7] Nakamura, L. K. 1989. Taxonomic relationship of black-pigmented *Bacillus subtilis* strains and a proposal for *Bacillus atrophaeus sp. nov*. International Journal of Systematic Bacteriology **39**:295-300.

Radiation Inactivation of Bioterrorism Agents
L.G. Gazsó and C.C. Ponta (Eds.)
IOS Press, 2005

Inactivation of Biological Threat Agents with Nonionizing Radiation

Gregory B. KNUDSON, Michael O. SHOEMAKER and Thomas B. ELLIOTT
Armed Forces Radiobiology Research Institute
8901 Wisconsin Avenue, Bethesda, Maryland 20889, USA
TEL: (301) 295-1947, FAX: (301) 395-6503

Abstract. The terrorist attack in September 2001, with the dissemination of *Bacillus anthracis* spores in letters sent through the U.S. Postal Service, brought home the reality of bioterrorism. These attacks have heightened concerns about future large-scale aerosol attacks with powders of *B. anthracis* spores and other pathogens that cause smallpox, pneumonic plague, tularemia, and viral hemorrhagic fevers, as well as toxins such as botulinum toxin, ricin or *Staphylococcus* enterotoxin B. Means to prevent the use of these agents, and to manage the consequences of their use, have become a high priority. One of the tools that can play a role in disease prevention and consequence management is nonionizing radiation in the form of germicidal short-wavelength ultraviolet (UV) light. This article presents background information on the pathogen *Bacillus anthracis*, the causative agent for anthrax, and its susceptibility to killing by germicidal UV. The results of two experimental studies are also presented that examine UV inactivation of *B. anthracis* vegetative cells and spores, and spores of closely related *Bacillus* species, in suspension, dried on surfaces, and as free-flowing powders.

Introduction

When *Bacillus anthracis* vegetative cells are nutritionally stressed, they initiate a complex sequence of events leading to sporulation [1]. The highly stable *B. anthracis* endospore, or henceforth spore, is the infectious particle resulting in anthrax. The form of the disease (cutaneous, gastrointestinal, or inhalational) is dependent on the route of spore entry into the host [2]. The spore consists of an exosporium surrounding a spore coat comprised mainly of protein and occupying about 50 percent of the spore volume. This coat encloses a peptidoglycan cortex and the spore core, which contains bacterial DNA complexed with small acid-soluble spore proteins and high concentrations of dipicolinic acid (DPA). The calcium ion chelate of DPA constitutes approximately 10 percent of *B. subtilis* spore dry weight and plays a significant role in spore resistance to ultraviolet (UV) radiation [3-6].

Developing improved methods for decontaminating personnel, equipment, air, water, food, buildings, and the environment following a bioterrorist attack with persistent biological agents is a high priority [7., 8]. In order to effectively integrate germicidal UV lights and solar UV irradiation into these decontamination processes, the biological effects and limitations need to be clearly understood. This paper describes the biophysics of UV inactivation of pathogens and DNA repair and presents two sets of studies that investigate the effectiveness of germicidal UV for inactivation of spores of *B. anthracis* and other *Bacillus* species.

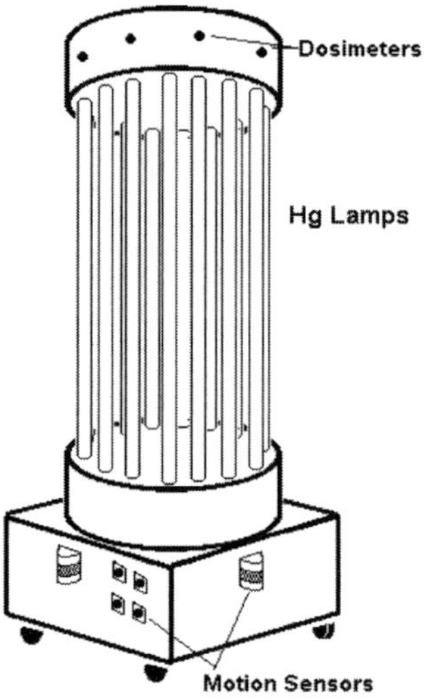

Figure 1. High-dose UV-C sterilization system tested at AFRRI for inactivating *B. anthracis* Sterne and *B. atrophaeus* (*B. subtilis* var. *niger*; *B. globigii*, "BG") spores on non-reflective surfaces. This mobile, UV area, decontamination device, invented and developed by Jeffery L. Deal, M.D. and D. Deal (UVAS- LLC, Charleston, SC), incorporates a bank of fourteen medium-pressure mercury-vapor tubes to generate intense levels of UV-C with a power output of approximately 3,000 microwatts/cm^2 at one meter.

1. UV-C Inactivation of *B. anthracis* and *B. atrophaeus* Spores Dried on Painted Metal Plates and Dry *B. atrophaeus* Spore Powder

The authors, in collaboration with Jeffery L. Deal and D. Deal (UVAS-LLC, Charleston, SC), Marie U. Owens (Department of Biology, College of Charleston, Charleston, SC) and Janet E. Meszaros (Steris Corporation, Mentor, OH), conducted an assessment of a commercially available UV area sterilization device. This mobile room decontamination device, invented by J. L. Deal and D. Deal, incorporated a bank of 14 medium pressure mercury vapor tubes to generate intense levels of ultraviolet-C (UV-C), with a power output of approximately 3,000 microwatts/cm^2 at one meter (Figure 1). An array of sensors on the unit detected the levels of UV reflected back to the unit. A central processing unit calculated the time required to deliver a preselected inactivating dose to the darkest area of the room. To assure personnel safety, the device has motion detectors that instantly shut down the unit if motion occurs in the room during irradiation.

In this study performed at the Armed Forces Radiobiology Research Institute (AFRRI) in Bethesda, MD, we examined the effectiveness of high-dose UV-C light for inactivating *B. anthracis* Sterne spores and *B. atrophaeus* ATCC 9372 spores (formerly *B. subtilis* var. *niger* and *B. globigii*, or "BG") suspended in 50 percent ethanol and dried on non-reflective metal surfaces, and thin films of dry, free-flowing *B. atrophaeus* spore powder 93-PBA-1 on reflective plastic surfaces. The latter, *B. atrophaeus* 93-PBA-1, contained approximately one percent silica, with an initial spore concentration of 2.5 x 10^{11} colony-forming units (CFU)/g.

1.1. UV Irradiation of Painted Metal Plates Spread with an Alcohol Spore Suspension

B. atrophaeus ATCC 9372 spores (Steris Corporation, Mentor, OH) were suspended in 50 percent ethanol and used at concentrations of 10^9, 10^5, and 10^4 CFU/ml. *B. anthracis* Sterne

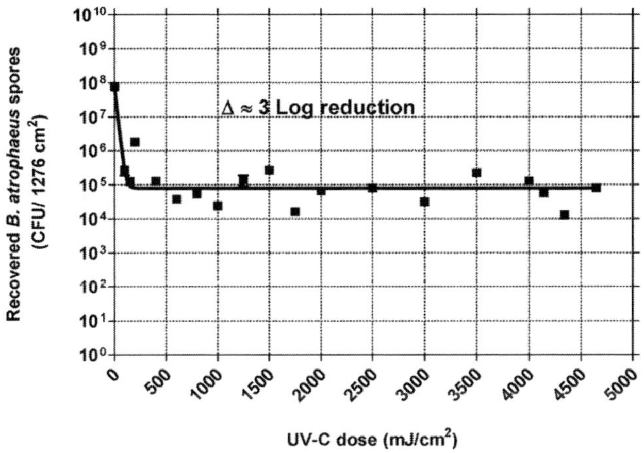

Figure 2. UV-C (254 nm) inactivation of *B. atrophaeus* ATCC9372 spores air dried on a flat-black metal surface and enumerated using the Millipore Swab Dilution Sampler recovery method. $D_{10} \approx 33$ mJ/cm^2 = 330 J/m^2 for initial log reduction.

spores were produced at AFRRI from a live-spore veterinary vaccine (Colorado Serum, Denver, CO) and suspended in 50 percent ethanol at a concentration of 2.3 x 10^9 CFU/ml. Using sterile plastic-rod spreaders ("hockey sticks"), 2-ml aliquots of these 50 percent ethanol-spore suspensions were distributed on test surfaces consisting of aluminum plates with surface areas of 1276 cm^2 painted with high heat-stable flat-black enamel and sterilized by autoclaving. An additional 5 ml of 50 percent ethanol was used to facilitate the even distribution of spore suspension over the entire surface area. Test surfaces were air dried for 2-3 hours before UV-C exposure. These dark surfaces were intended to represent the worst conditions for ultraviolet light reflectance.

Two methods for recovering spores from the metal plates following UV irradiation were tested. At time zero and after various UV exposures, 45-mm contact culture plates containing trypticase soy agar with lecithin and polysorbate 80 (Rodac plates, PML Microbiologicals) were used to recover viable spores of *B. anthracis* Sterne from the metal surfaces. Following each UV exposure time, Swab Dilution Samplers (Millipore Corporation, MT0010025) were also used to sample the metal surfaces for recovery of *B. atrophaeus* and *B. anthracis* Sterne. According to the manufacturer's instruction, a 16-cm^2 area of the painted test surface was swabbed and spores were eluted in the Millipore buffer. The suspension was serially diluted and 150-µl aliquots were plated in triplicate on 90-mm trypticase soy agar (TSA) plates (BD Diagnostic Systems). The Rodac and TSA plates had been incubated at 35°C for 15 hours when the colonies were counted. Knowing the surface area sampled and the dilution and volume plated, the plate counts were used to compute the number of viable spores remaining on the total metal surface (1276 cm^2) following each UV exposure.

Colony counts from the Millipore Swab Dilution Samplers showed that at time zero, 7.7 x 10^7 CFU/1276 cm^2 could be recovered from the metal plate that had been spread with *B. atrophaeus* spores. After a UV dose, or energy fluence, of 500 mJ/cm^2, there was a 3-log reduction in recoverable viable *B. atrophaeus* spores to 8 x 10^4 CFU/1276 cm^2. Following this initial 3-log drop in viable spores, there was no additional spore killing with increasing doses up to 4,000 mJ/cm^2 (Figure 2).

Using the same Millipore swab sampling method to recover and enumerate viable spores from metal plates spread with a suspension of *B. anthracis* Sterne spores, there were 3.6 x 10^8 CFU/1276 cm^2 just prior to UV exposure. The spore count was reduced to ap-

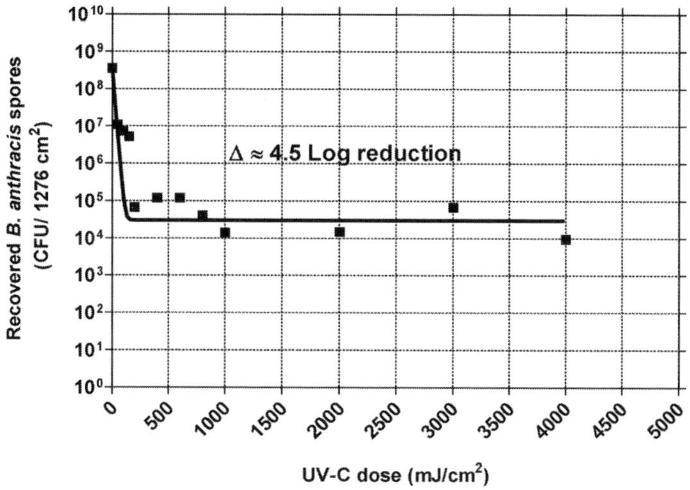

Figure 3. UV-C (254 nm) inactivation of *B. anthracis* Sterne spores air-dried on a flat-black metal surface and enumerated using the Millipore Swab Dilution Sampler method. An initial 4.5-log reduction in recoverable viable spores was observed and was followed by "tailing" where no additional killing of the spores occurred with increasing doses of UV-C. $D_{10} \approx 22$ mJ/cm^2 = 220 J/m^2 for the initial log reduction.

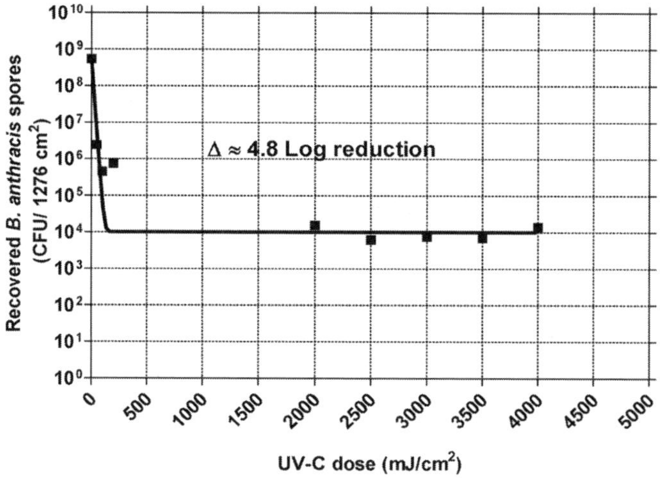

Figure 4. UV-C (254 nm) inactivation of *B. anthracis* Sterne spores air dried on a flat-black metal surface and enumerated using the Rodac plate recover method. Following an initial 4.8-log reduction in the viability counts, "tailing" occurred in which no additional killing of the spores was observed with increasing doses of UV-C. $D_{10} \approx 21$ mJ/cm^2 = 210 J/m^2 for the initial log reduction.

proximately 1 x 10^4 CFU/1276 cm^2 following a UV exposure of 500 mJ/cm^2. After an initial 4.5-log reduction in the viable counts, a "tailing" of the survival curve occurred where no additional killing of spores was observed with increasing doses of UV-C up to 4,000 mJ/cm^2 (Figure 3). Using the Rodac contact plate method, the recovery of viable *B. anthracis* spores was plotted as a function of UV dose. There was an initial 4.8-log reduction in the viable spore counts on Rodac plates followed by "tailing" of the curve where no additional killing was observed with increasing doses of UV-C up to 4,000 mJ/cm^2 (Figure 4).

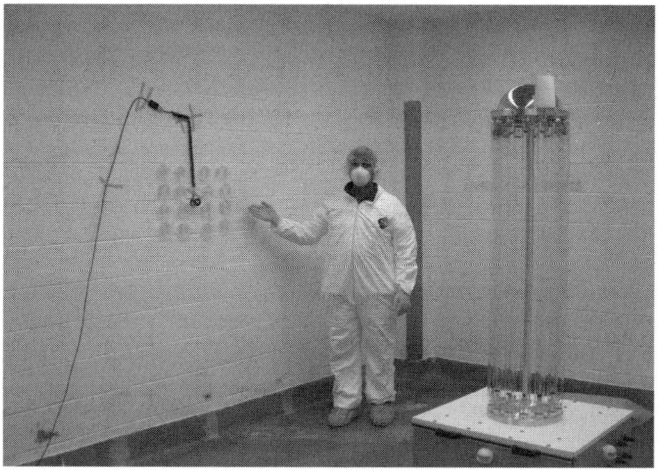

Figure 5. Experimental setup for inactivation of dry free-flowing *B. atrophaeus*; 93-PBA-1 spores. Petri dishes containing spore powder were attached to the wall with tape in the 25- x 35-ft UV exposure room at AFRRI. A calibrated UV dosimeter (National Institute of Standards and Technology) was mounted on the wall in the center of the array of 16 plates. The mobile sterilization device is shown on the right.

1.2. UV Irradiation of Open Petri Plates Containing a Thin Film of Dry Spore Powder

Approximately 35 mg of *B. atrophaeus* 93-PBA-1 was spread on sterile 90-mm Petri plates by adding a uniform dry measure of 0.5 cc spore powder and gently rotating the plates by hand to achieve a relatively uniform distribution of the powder. Excess spore powder was removed by inverting and tapping the plate gently. To reduce shadowing effects during UV-C exposure, powder was wiped from the sides and corners of the Petri dishes with sterile cotton swabs. The mass of spores remaining in each plate was determined by weighing each plate before and after the addition of spore powder; the *B. atrophaeus* spore powder averaged 0.039 g/plate ± 0.008 SD (range 0.023 to 0.049 g/plate).

The *B. atrophaeus* spore-powder Petri plates were affixed to a wall directly in front of the UV-C source, and the Petri plate lids were removed just prior to UV irradiation. A calibrated UV dosimeter (National Institute of Standards and Technology, Gaithersburg, MD) was mounted on the wall in the center of the array of plates (Figure 5). After delivery of each UV radiation dose, a plate was removed from the exposure wall, and the spores were harvested by washing the plates three times using a total of 5 ml sterile water containing 0.05 percent Tween 80. A sterile rubber cell spreader was used to aid in removing the spore powder from the plates. The spore suspensions were vortexed and diluted, and 150-µl aliquots of each dilution were inoculated onto TSA plates in triplicate. The cultures were incubated at 35°C for 15 hours at which time colonies were counted.

Irradiated Petri dishes containing *B. atrophaeus* spore powder showed little inactivation even at very high doses of UV. There was an initial reduction in viability of approximately one log followed by "tailing" in which no additional inactivation was achieved even at doses as high as 16,000 mJ/cm^2 (Figure 6). Because of the poor penetrating power of UV radiation, shadowing from irregularity of the surface and self-shielding, or clumping of spores, may have provided protection against the biocidal action of the UV radiation.

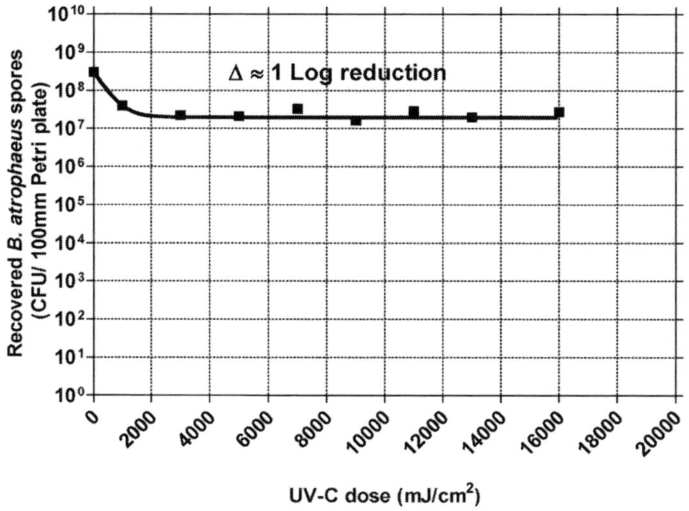

Figure 6. UV-C (254 nm) inactivation of dry, free-flowing *B. atrophaeus* spore powder 93-PBA-1 containing 1 percent silica. After each UV exposure, plates containing dry spore powder were removed form the exposure wall and washed with sterile water containing 0.05 percent Tween 80, and the dilutions of the suspension were inoculated onto TSA plates. The UV inactivation curve shows an initial 1-log reduction in the viable spore counts followed by "tailing" of the curve where no additional killing of the spores was observed with increasing doses of UV-C.

2. UV-C Inactivation of Virulent *B. anthracis* Spores and Vegetative Cells

Bacillus species have considerable diversity with regard to their ability to repair UV damage through photoreactivation. Vegetative cells of *B. mycoides*, *B. pumilus*, *B. megaterium*, and some strains of *B. cereus* can be photoreactivated while *B. subtilis*, *B. polymyxa*, and the *B. circulans* complex lack photoreactivation [9]. One of the authors of this paper (G. B. Knudson) established the kinetics of UV inactivation of vegetative cells of various strains of *B. anthracis* with and without the plasmids pXO1 and pXO2 and *B. anthracis* endospores. This research [10] also demonstrated that vegetative cells of *B. anthracis* are capable of photoreactivation.

The *B. anthracis* UV survival curves published by Knudson [10] were used to calculate that the UV fluence required to reduce the initial population of *B. anthracis* Sterne spores by 90 percent (LD_{90}) is approximately 810 J/m^2. They also showed that, from the log-linear portion of the curve, the UV fluence corresponding to the decimal reduction value (D_{10}) is approximately 540 J/m^2 whereas the D_{10} of vegetative cells of *B. anthracis* Vollum is much lower at approximately 14 J/m^2 (Figures 7 and 8). The source of UV in these studies was a 15-W G15T8 germicidal lamp (General Electric Co., Schenectady, NY), which emits mainly at 254 nm, and the flux was 0.9 W/m^2, as measured with a shortwave UV meter (model J-225; Ultra-Violet Products Inc., San Gabriel, CA). Spore resistance to UV radiation is greatly influenced by the specific conditions of spore formation. Under different spore preparation and irradiation conditions, a D_{10} value for *B. anthracis* Sterne spores is reported to be as low as 275 J/m^2 [11]. UV inactivation values for bacterial spores are dependent on a number of variables, including irradiation conditions, power and wavelength of the UV source, bacterial species and repair efficiency, growth and sporulation conditions, and spore preparation methods.

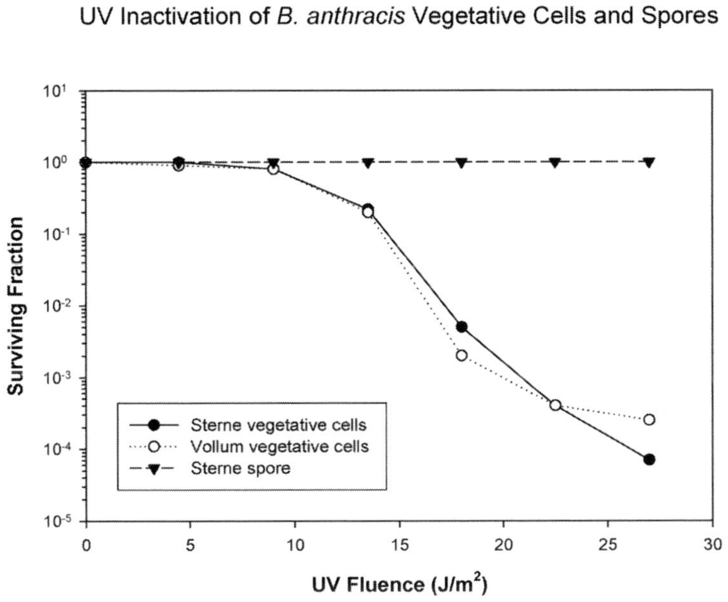

Figure 7. UV-C (254 nm) inactivation of virulent vegetative *B. anthracis* Vollum cells, vegetative *B. anthracis* Sterne cells, and *B. anthracis* Sterne spores. The survival curve for the virulent Vollum strain is not significantly different from the survival curve for the avirulent Sterne strain. The Vollum strain has pXO1 (toxin-encoding plasmid) and pXO2 (poly-D-glutamic-acid-encoding plasmid) plasmids whereas the Sterne strain has a pXO1 but lacks the pXO2 plasmid. At UV doses less than 30 J/m^2, no loss in *B. anthracis* Sterne spore viability was observed. $D_{10} \approx 14$ J/m^2 for the initial log reduction of *B. anthracis* vegetative cells.

3. Discussion

3.1. Distinction between Ionizing and Nonionizing Radiation-induced Cellular Injuries

UV light is generally considered to be electromagnetic energy with wavelengths between 190 and 380 nm, although for some purposes it has been defined as wavelengths extending from 100 to 400 nm. In biophysical literature, UV is often subdivided into near-UV (300-380 nm) and far-UV (190-300 nm) while in biomedical literature, it is often arbitrarily subdivided into UV-A (320-380 nm), UV-B (290-320 nm), and UV-C (190-290 nm). High intensity, low-pressure, mercury-vapor discharge lamps (germicidal lamps) used for killing microorganisms in air and water emit 97.4 percent of their far UV emission at 253.7 nm [12].

UV radiation, and all less energetic radiation including visible light, is considered to be nonionizing. Radiation with shorter wavelengths such as x-rays and gamma-rays and high-speed particles such as alpha and beta particles are referred to as ionizing radiation. Absorption of nonionizing radiation can lead to electronic excitation of organic molecules whereas ionizing radiation is energetic enough to dissociate electrons in organic molecules and can cause cell damage by generating highly reactive hydroxyl radicals. The target molecules and the damage caused by the different types of radiation in the cell are therefore different. Since the DNA molecule absorbs radiation in the UV range with a peak absorption at 260 nm, the most profound effects on the cell are due to unrepaired, or missrepaired, DNA. Ionizing radiation can damage cellular DNA by causing single or double strand breaks, crosslinking, or base substitutes resulting in mutational effects or cell death. Ionizing radia-

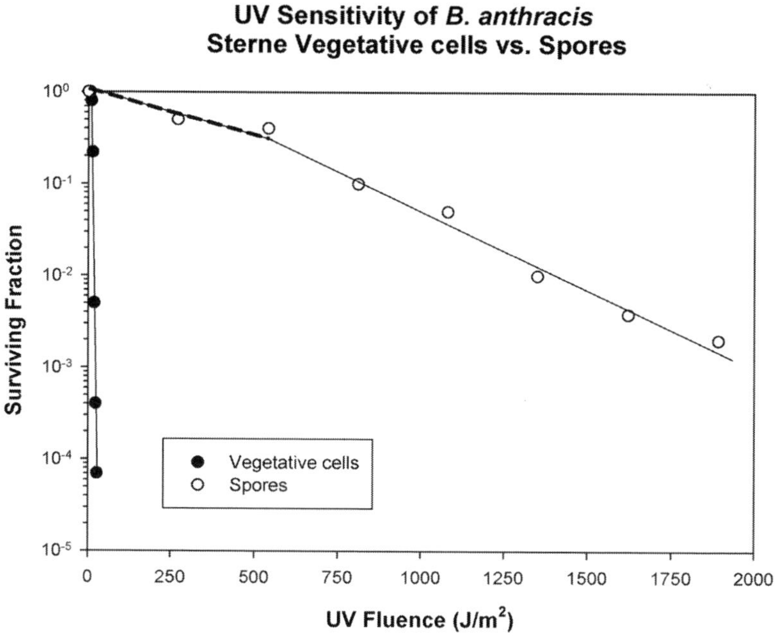

Figure 8. UV-C sensitivity of *B. anthracis* Sterne vegetative cells versus spores. At UV-C doses between 250 and 1900 J/m^2, log-linear inactivation of *B. anthracis* Sterne spores was observed. The comparative susceptibility of *B. anthracis* vegetative cells to UV-C radiation is much greater than that of spores. $D_{10} \approx 810$ J/m^2 for the initial log reduction of *B. anthracis* Sterne spores.

tion can also damage other critical molecules in the cell resulting in the cell's inability to replicate.

3.2. Units of Measure

Energy fluence (previously called UV dose) applied to the biological target is commonly expressed in J/m^2. One J/m^2 is equal to 0.1 mJ/cm^2 or 10^3 erg/cm^2 or 100 μW·sec/cm^2. The fluence can be calculated by multiplying the fluence rate (flux as measured by a photovoltaic meter) by the total time of exposure. For example, if the UV lamp is set at a distance that gives a flux of 0.9 W/m^2, then an exposure time of 15 seconds would give a fluence of 90 μW/cm^2 x 15 sec = 1350 μW·sec/cm^2 = 135 erg/mm^2 = 13.5 J/m^2.

The absorption spectrum for DNA has an absorption maximum in the 260-265 nm range. This DNA absorption spectrum correlates with the inactivation efficiency as a function of wavelength. This correlation between the DNA absorption spectrum and the bactericidal spectrum was first reported by F. L. Gates [13] and has been developed into a UV inactivation model for *B. anthracis* Ames spores [14].

3.3. UV Photoproducts and Cell Inactivation

The effect of pryrimidine dimers on cell survival and repair as well as the mechanisms by which UV-induced cyclobutyl pyrimidine dimers result in mutation have been extensively studied [15,16]. Pyrimidine dimers, considered the most important UV photoproducts formed in DNA molecules, are formed between two adjacent pyrimidines in the same stand of DNA resulting in local denaturation of the DNA helix. Photochemical alterations in

DNA following UV exposure of vegetative cells are primarily manifested by the formation of pyrimidine dimers (5,6-cyclobutyl dipyrimidines). Thymine-thymine dimers are more frequent than thymine-cytosine and cytosine-cytosine dimers. During DNA replication, a non-repaired thymine dimer can be a lethal event because DNA polymerase will terminate the synthesis at that point. Since there is no template at the point of the dimer, DNA polymerase will leave a gap and continue synthesizing the new strand further downstream of the dimer.

In contrast to the cyclobutane-type of thymine dimers formed in DNA found in vegetative bacteria, the primary UV-induced photoproduct formed in DNA found in spores of *Bacillus* species has been identified as the thymine dimer, 5-thyminyl-5,6-dihydrothymine. The spore photoproducts are not cleaved by the photoreactivation enzyme, but the accumulated spore photoproducts can be repaired in *B. subtilis* spores by several dark repair systems during germination [17, 18].

UV-induced bacteria inactivation is observed as the inability to replicate and form a visible colony. A decimal-reduction curve can be constructed by plotting the surviving fraction of the total initial number of spores on a logarithmic scale in terms of viable CFU as a function of the UV fluence. Such survival curves for vegetative bacterial cells often show an initial “shoulder” at low UV fluences followed by exponential inactivation and a declining rate of inactivation, or “tailing” effect. This "shoulder" may be a result of the requirement of multiple hits for inactivation, or the result of the fact that at low doses of UV, DNA repair enzymes can effectively counter the damage through photoreactivation, excision repair, postreplication recombinational repair, SOS error-prone repair, and other repair processes. Mutant strains of bacteria deficient in excision repair such as *Escherichia coli* B_{s-1} are much more sensitive to the inactivating effects of UV radiation. Survival curves that end in a more UV-resistant tail may reflect population heterogeneity with a subpopulation capable of more effective DNA repair [19., 20]. Tailing in our survival curves for UV-irradiated spore powder is most likely the result of shielding within micrometer-sized clusters. This type of curve can be represented by two logarithmic functions: one with a rate constant for the initial rapid inactivation and one with a smaller rate constant for the tailing. These rate constants, or decimal-reduction values (D_{10}), are the measured value of the inactivating agent required to decrease the number of infectious agents by one logarithm.

3.4. UV-induced Mutagenesis and DNA Repair

Various DNA repair processes can repair pyrimidine dimers prior to DNA replication (e.g., photoreactivation, excision repair) or repair the gaps following replication (e.g., recombinational repair). The gaps can also be filled by a process called Weigle reactivation, which leads to increased cell survival but also to increased mutation rates.

Photoreactivation is the enzymatic reversal of UV-induced cyclobutane pyrimidine dimers through *in situ* monomerization in the presence of visible light (300-600 nm). This UV repair process was first reported by A. Kelner [21] and R. Dulbecco [22] and is now recognized to be widely distributed among both prokaryotic and eukaryotic organisms. Although numerous bacteria possess photoenzymatic repair capabilities, several members of the genus *Bacillus* lack photoreactivation. The photoreactivating enzyme binds to pyridine dimers, which distort the double helix. Then, with the addition of photoreactivating light-energy wavelengths between 310 and 480 nm, the photoreactivating enzyme cleaves the dimer, thus restoring the two pyrimidines *in situ*. A short flash of requisite illumination with visible light is sufficient to photolyze enzyme-substrate complexes. *E. coli* B contains approximately 20 photoreactivating enzyme molecules per cell [17].

Excision repair, or “cut-and-patch,” is the enzymatic process of excising an area of one DNA strand that contains a pyrimidine dimer with concurrent filling of the gap by repair synthesis using the complimentary strand as a template followed by strand ligation [23, 24].

Postreplication recombinational repair is a process of filling gaps in a daughter strand, a strand formed during DNA synthesis opposite pyrimidine dimers in the parental strand. When DNA synthesis is blocked by a pyrimidine dimer, DNA polymerase will leave a gap of approximately 1,000 nucleotides before continuing to synthesize the new daughter strand. In wild type cells (with Rec A and Lex A proteins), these postreplication gaps can be filled with nucleotide sequences from the homologous parental strand through a process of recombination. The remaining pyrimidine dimers can then be repaired by excision repair or photoreactivation.

The DNA repair processes described above have very high fidelity. An error-prone repair process (variably referred to as Weigle reactivation, W-reactivation, UV-inducible error-prone repair, or SOS repair) was first described by J. J. Weigle [25., 26]. Weigle's initial observation discovered that the survival of UV-irradiated bacteriophage was higher when the infected host cells had also been previously irradiated, and the surviving phage had a higher mutation rate. This UV-inducible error-prone repair system has also been observed with animal viruses infecting irradiated animal cells. The likely mechanism is that when the host cells are irradiated and post replication gaps form, the repair system is induced to directly fill the gaps. Since a pyrimidine dimer does not present any bases to pair against, the repair process is error-prone. Filling gaps in this manner increases both survival and the mutation rate. When previously UV-inactivated viruses are used to infect host cells with induced error-prone repair enzymes, the host cell will repair the virus DNA as well as the cellular DNA. The repaired virus can then replicate and kill the host cell in which it was repaired.

Biotechnology techniques could be used to modify the DNA of threat agents to enhance their resistance to antibiotics and vaccines, increase their virulence and pathogenicity, modify their host range, and increase their survivability in adverse environmental conditions including prolonged exposure to solar UV. By enhancing or adding DNA repair genes, it is possible that a genetically engineered pathogen could survive longer after release, thus increasing the probability of infecting a host.

3.5. Factors Affecting UV Inactivation

The factors affecting the UV dose required to sterilize an object contaminated with bacterial spores include the following: the wavelength of UV, the initial number of bacterial spores or bioburden prior to irradiation, the species of *Bacillus* or *Clostridium*, the DNA repair competence, the process by which the spores were produced, the degree of spore hydration, the gaseous environment in which the spores were irradiated, the temperature during irradiation, the presence of photosensitizers or protectors such as blood and serum, and the amount of shielding present.

3.6. Solar UV and Countermeasures

The stratospheric ozone layer effectively absorbs the most energetic UV wavelengths, namely UV-C and most UV-B. Biologically relevant solar UV with wavelengths as short as 300 nm can reach the earth's surface and interact with cellular DNA, causing the formation of thymine dimers. The solar UV index at the Earth's surface is dependent on a number of factors including time of day and year, latitude, cloud cover, and ground reflection. Sunlight also contains longer wavelength photons required for photoreactivation that repair the pyrimidine dimers formed in DNA when UV light is absorbed.

B. anthracis spores are remarkably resistant to UV inactivation compared to vegetative cells. Bacterial and viral pathogens purposely disseminated as an aerosol cloud are more vulnerable to solar UV inactivation than pathogens that have settled out onto irregular surfaces. However, the greatest health risk comes from inhalation of the initial aerosol as opposed to infection from fomites or from secondary aerosols. To avoid a solar UV inactivation decrement, a pathogen aerosol would likely need to be disseminated in the early morning or evening, which also corresponds with the ideal time for air inversions to keep the aerosol cloud close to the ground. Since the shorter wavelength UV increases by about 6 percent with each 1 km increase in altitude, the premature release of a biological warfare agent in the upper atmosphere, as a result of ballistic missile interception for example, would expose the cloud to a larger fluence of germicidal UV than if the cloud was released at ground level.

Microencapsulation could also increase the UV protection and retard dehydration effects for bacterial and viral pathogens. In developing additives to increase the persistence of viral insecticides, Tamex-Guerra, et al. [27] showed that a spray-dried virus embedded within microgranules of corn flour and potassium lignite had significantly improved activity after prolonged exposure to sunlight. Furthermore, Vettori, et al. [28] reported that clay-associated bacteriophages are protected against UV radiation inactivation compared to free phage not adsorbed on clay minerals such as kaolinite. The persistence of baculoviruses in the environment has been increased through microencapsulation with UV screening of carbon black or titanium dioxide [29]. Large quantities of African and Asian dust, consisting primarily of clay soil minerals such as kaolinite transport viable microorganisms to the Americas [30]. Bacteria and fungi probably remain viable during long-range transport at high elevations as a result of UV light attenuation by the dust clouds and through shielding within cracks and crevices of the clay particles.

4. Conclusions

4.1. Use and Limitations of UV for Decontamination of Air, Water, and Surfaces

UV-irradiation-based processes are effective in the disinfection of water, surfaces, and air that are contaminated with most microorganisms. For example, germicidal UV (254 nm) is a feasible modality for the control of Legionnaires' disease bacterium in water. G. B. Knudson [16] published UV dose-survival curves for *Legionella pneumophila* and six other *Legionella* species [16], which were very sensitive to low doses of UV (e.g., 240 $\mu W/cm^2$ for 30 seconds) and also had strong DNA repair photoreactivation capabilities. UV treatment combined with water filtration is effective in preventing *Legionella* colonization in hospital water fixtures [31]. Because of the limited penetrating power of UV radiation, UV disinfection of turbid water is more effective when pretreated to remove particulate matter, organic compounds, and colored compounds. Germicidal UV lights also are successfully used in decontamination systems for reducing the number of airborne pathogens in air. For example, UV germicidal irradiation can be effective in controlling airborne dissemination and transmission of *Mycobacterium tuberculosis* in hospitals [32]. Commercially available germicidal UV irradiation equipment can be used in the upper part of occupied spaces and within ductwork of large buildings along with high efficiency filters and dilution ventilation strategies to protect against the intentional release of bioterrorism or biowarfare pathogens [33].

It is important to understand the capabilities of UV disinfection systems as well as their intrinsic limitations [12]. Unlike ionizing radiation, UV light has very limited penetrating power. Our current studies demonstrate that spores of biological warfare agent surrogates dried on irregular surfaces will not be inactivated if the spores are shielded from the UV light in microscopic scratches, cracks, and crevices. We also demonstrated that visible spore powders on surfaces are not effectively inactivated even at very high UV fluences perhaps because of the self-shielding of spores within the powder.

4.2. Need for Additional Studies

Further research is needed to assess countermeasure applications for germicidal UV in decontaminating air, water, and surfaces contaminated with pathogens and protein toxins. In particular, the effectiveness of UV-protective microencapsulation needs to be assessed. Intrinsic characteristics and conditions for the use of UV sources designed for field applications need to be evaluated and quantified, including UV penetration under field conditions, shielding problems due to agent clumping, and obstruction due to battlefield smoke. The development of nontoxic photosensitizing dyes that couple with biological agents and toxins to enhance photoinactivation could improve practical countermeasure applications. Photochemical inactivation of biological agents combined with photsensitizers could be induced by sunlight or low-photon energy sources such as high-temperature UV-emitting flares and lasers.

References

[1] Liu, H., N.H. Bergman, B. Thomason, S. Shallom, A. Hazen, J. Crossno, D. A. Rasko, J. Ravel, T.D. Read, S.N. Peterson, J. Yates III, and P.C. Hanna. 2004. Formation and composition of the Bacillus anthracis endospore. Journal of Bacteriology, 186:164-178.

[2] Knudson, G.B. 1986. Treatment of anthrax in man: Historical and current concepts. Military Medicine, 151:71-77.

[3] Setlow, P. 1994. Mechanisms which contribute to the long-term survival of spores of *Bacillus* species. Journal of Applied Bacteriology, Symposium Supplement, 76: 495-605.

[4] Slieman, T.A. and W. L. Nicholson. 2001. Role of dipicolinic acid in survival of *Bacillus subtilis* spores exposed to artificial and solar UV radiation. Applied and Environmental Microbiology, 67:1274-1279.

[5] Nicholson, W. L., N. Munakata, G. Horneck, H. J. Melosh, and P. Setlow. 2000. Resistance of *Bacillus* endospores to extreme terrestrial and extraterrestrial environments. Microbiology and Molecular Biology Reviews, 64:548-572.

[6] Riesenman, P. J. and W. L. Nicholson. 2000. Role of the spore coat layers in *Bacillus subtilis* spore resistance to hydrogen peroxide, artificial UV-C, UV-B, and solar UV radiation. Applied and Environmental Microbiology, 66:620-626.

[7] Knudson, G. B. 1991. Operation Desert Shield: Medical Aspects of Weapons of Mass Destruction. Military Medicine 156:267-271.

[8] Knudson, G. B. 2001. Nuclear, biological, and chemical training in the U.S. Army Reserves: Mitigating psychological consequences of weapons of mass destruction. International Conference on the Operational Impact of Psychological Casualties from Weapons of Mass Destruction. Military Medicine 166(12):63-65.

[9] Russell, A. D. 1982. The Destruction of Bacterial Spores. New York: Academic Press.

[10] Knudson, G.B. 1986. Photoreactivation of ultraviolet-irradiated, plasmid-bearing, and plasmid-free strains of *Bacillus anthracis*. Applied and Environmental Microbiology, 52:444-449.

[11] Nicholson, W.L., and B. Galeano. 2003. UV resistance of *Bacillus anthracis* spores revisited: Validation of *Bacillus subtilis* spores as UV surrogates for spores of *B. anthracis* Sterne. Applied and Environmental Microbiology, 69:1327-1330.

[12] Morris, E. J. 1972. The practical use of ultraviolet radiation for disinfection purposes. Medical Laboratory Technology, 29:41-47.

[13] Gates, F.L. 1928. On nuclear derivatives and the lethal action of ultraviolet light. Science, 68:479-480.

[14] Seely, S.L., and J. M. Shuford. 2002. Modeling inactivaiton of *B. anthracis* by ultraviiolet radiation. Third Symposium on Environmental Applications, Orlando, FL. Session 6, 6.2-6.3.

[15] Sinha, R. P., and D. P. Hader. 2002. UV-induced DNA damage and repair: A review. Photochemical and Photobiological Science, 1(4):225-236.

[16] Knudson, G. B. 1985. Photoreactivation of UV-irradiated *Legionella pneumophila* and other *Legionella* species. Applied Environmental Microbiology 49:975-980.

[17] Harm, W. 1980. Biological effects of ultraviolet radiation. Cambridge, NY: Cambridge University Press.

[18] Setlow, P. 1995. Mechanisms for the prevention of damage to DNA in spores of *Bacillus* species. Annual Review of Microbiology, 49:29-54.

[19] Cerf, O. 1977. A Review: Tailing of survival curves of bacterial spores. Journal of Applied Bacterioogy, 42:1-19.

[20] Blatchley, E.R. III, and M.M. Peel. 2001. Disinfection by ultraviolet irradiation. In: Disinfection, Sterilization, and Preservation, *ed.* S.S. Block. Lippincott Williams & Wilkins, Philadelphia, pp.1023-1047.
[21] Kelner, A. 1949. Photoreactivation of UV-irradiated *Escherichia coli* with special reference to the dose-reduction principle and to UV-induced mutation. Journal of Bacteriology, 58:511-522.
[22] Dulbecco, R. 1949. Reactivation of ultraviolet inactivated bacteriophage by visible light. Nature, 163:949-950.
[23] Boyce, R.P., and P. Howard-Flanders. 1964. Release of ultraviolet light-induced thymine dimers from DNA in *E. coli* K12. Proceedings of the National Academy of Science U.S.A., 51:293-300.
[24] Setlow, R.B. and W.L. Carrier. 1964. The disappearance of thymine dimmers from DNA. Proceedings of the National Academy of Science U.S.A., 51:226-231.
[25] Weigle, J. J. 1953. Induction of mutation in a bacterial virus. Proceedings of the National Academy of Science U.S.A., 39:628-636.
[26] Knudson, G. B. 1983. The role of inducible DNA repair in W-reactivation and related phenomena. In: FE Hahn (ed) Progress in Molecular and Subcellular Biology, Vol. 8, pp. 22-40. Berlin: Springer-Verlag.
[27] Tamez-Guerra, P., M. R. McGuire, R. W. Behle, J. J. Hamm, R. R. Sumner, and B. S. Shasha. 1999. Sunlight persistence and rainfastness of spray-dried formulations of Baculovirus isolated from *Anagrapha falcifera* (Lepidoptera: Noctuidae). Journal of Economic Entomology, 93:210-218.
[28] Vettori, C., E. Gallori, and G. Stotzky. 2000. Clay minerals protect bacteriophage PBS1 of *Bacillus subtilis* against inactivation and loss of transducing ability by UV radiation. Canadian Journal of Microbiology, 46:770-773.
[29] Ignoffo, C. M., and O. F. Batzer. 1971. Microencapsulation and ultraviolet protectants to increase sunlight stability of an insect virus. Journal of Economic Entomology, 64:850-853.
[30] Garrison, V. H., E. A. Shinn, W. T. Foreman, D. W. Griffin, C. W. Holmes, C. A. Kellogg, M. S. Majewski, L. L. Richardson, K. B. Ritchie, and G. W. Smith. 2003. African and Asian dust: From desert soils to coral reefs. BioScience, 53:469-480.
[31] Liu, Z., J.E. Stout, L. Tedesco, M. Boldin, C. Hwang, and V. L. Yu . 1995. Efficacy of ultravioltet light in preventing *Legionella* colonization of a hospital water system. Water Research, 29:2275-2280.
[32] Nardell, E.A. 1993. Environmental control of tuberculosis. Medical Clinics of North America. 77:1315-1334.
[33] Brickner, P.W., R. L. Vincent, M. First, E. Nardell, M. Murray, and W. Kaufman. 2003. The application of ultraviolet germicidal irradiation to control transmission of airborne disease: Bioterrorism countermeasure. Public Health Reports, 118:99-114.

Radiation Inactivation of Bioterrorism Agents
L.G. Gazsó and C.C. Ponta (Eds.)
IOS Press, 2005

Ionizing Radiation Inactivation of Medically Relevant Viruses

R. Joel LOWY
Armed Forces Radiobiology Research Institute
8901 Wisconsin Avenue, Bethesda, Maryland, 20889, USA

Abstract. Exposure of viral pathogens to ionizing radiation can be an effective method of neutralization. Potential applications include decontaminating materials and providing non-infectious antigens for diagnostic and therapeutic applications. As radiation can completely penetrate through most biological and non-biological materials, there can be high assurance that all the viruses have been completely exposed. This report will include recent experimental work investigating the radiation sensitivity of influenza A and vaccinia viruses, as well as a brief review of the published literature. Data to be presented includes the dose range for virus inactivation, comparisons of different types of ionizing radiation and putative molecular mechanisms. Biological and methodological parameters affecting radiation sensitivity and its measurement will also be discussed.

1. Introduction

1.1. Overview

Ionizing radiation potentially is an important method for neutralizing bio-warfare/bio-terrorism (BW/BT) viral agents. The focus of this manuscript is on medically relevant viruses, e.g. those that infect humans. Although radiation inactivation of viruses has been examined periodically from the middle of the last century, not all of this data is applicable to designing processes for putative BW/BT agent defeat. Also, despite a considerable body of work, there is as yet not a reliable theoretical framework to predict the relationship between radiation exposure and titer under a wide variety of conditions. Therefore, further empirical work is warranted. The intent of this paper is to provide a broad overview of the importance of further studies, a brief synthesis of what has been published, indicate where further work is needed and summarize some recent experimental work and comment on issues of experimental design.

1.2. Problems and Challenges Posed by Viral Agents

Viral diseases in the context of BW/BT are of great concern for a number of reasons. Even at present, most viral pathogens cannot be treated by a specific medical therapy and the most effective medical intervention is pre-exposure protection by vaccination. The problems with SARS (Severe Acute Respiratory Syndrome), West Nile Virus, Ebola and Human Immunodeficiency Virus (HIV) illustrate the difficulties in providing treatment beyond basic supportive therapy for viral diseases for which there are no vaccines. They also illustrate the problem for developing new vaccines even for viral diseases whose agents are accurately identified and well characterized. Influenza outbreaks, which kill approximately 35,000 US citizens annually, illustrate the severe epidemiological effects that can occur even with vaccination, in a medically sophisticated setting, with a long known and intensively studied viral pathogen. Human pathogens are not the only viruses of concern as viruses can also have devastating effects on food crops and animals. Recent outbreaks having profound economic and social effects include Nipha virus in Malaysia, the foot-and-mouth disease in Great Britain and the new avian strains of influenza virus in South-East Asia. These outbreaks affected three of the world's most important livestock species, respectively, swine, cattle, and poultry. Note that Nipha and the avian influenza viruses are also zoonotic, and caused human fatalities. In each case, attempts at quarantine were initially minimally successful and containment required destruction of large numbers of animals. Fortunately, there are increasing numbers of vaccines available or under development. Also, progress is being made on antivirals for some pathogens. Unfortunately, availability of new therapies to BW/BT threat agents have the potential of making their use more attractive to aggressors as it might be possible to protect themselves. Genetic engineering techniques make it increasingly possible to manipulate the genomes of viruses to produce chimeras with increased pathogenicity combined with better delivery and targeting to the species (generally human) of interest. The same recombinant DNA technology used to increase the speed of vaccine development has the potential to decrease or render ineffective those that we presently possess for any specific BW/BT agent.

1.3. Need for Neutralization and Decontamination

Therefore, it is likely that classic public health measures will remain an important part of the response to new emerging viral diseases and for response to BW/BT incidents. These include decontamination, sanitization, and sterilization that are important for the prevention and control of viral diseases. Such measures are routinely used in medical and research facilities, the medical and pharmaceutical manufacturing industries as well as normal public and personal hygiene. Commonly used microbicides include a wide range of chemicals, generally in liquid or gaseous form, wet and dry heat, non-ionization radiation, e.g. ultraviolet, and ionizing radiation [1, 2]. All these methods are used extensively and their applicability is dependent on the microbes to be inactivated, the material to be treated, and practical constraints, such as the speed needed for material handling. Microbial considerations include the type and virulence and the total amount of each infectious agent. Other issues include the volume, weight, surface area, shape complexity, quantity of the contaminated material, and the access the agent would have to all the contaminated surfaces [1,2]. Many of these issues are illustrated in several of the other manuscripts in this volume.

Such methods become even more necessary for the health of unprotected populations during outbreaks of infectious agents and are crucial during an epidemic. Treatment of a wide range of contaminated areas and materials as part of the response to an infectious disease outbreak is essential. Cleaning and infection control in medical facilities, of buildings, equipment, reusable and disposable supplies, are routinely performed and in some cases radiation is the method of choice. Purposeful contamination of public and private facilities is also possible, but likely much more problematic, as they are not designed for easy, thorough decontamination. Foot-and-mouth and avian flu, although not human diseases, provide the best and most recent examples of the difficulties, resources, and measures that may be needed for viral disease containment under field conditions during widespread epidemics. The actual measures needed would, of course, be much more complex and problematic to implement with a human epidemic. The recent problems with *Bacillus anthracis* spores within the United States postal system illustrates the wide range of material that can be contaminated and the difficulty of remediation. In the context of BW/BT a particularly difficult situation would be sealed containers containing highly virulent infectious agents at high titers. In situ inactivation methods of BW/BT agents at the production or deployment facilities would be highly advantageous.

2. Ionizing Radiation

2.1. Ionizing Radiation Overview

The major advantage ionizing radiation has over all other inactivation agents is its penetrating ability through materials. In addition, conditions necessary to deliver an exact dose throughout the target volume are well understood, as are the methods for measuring and verifying the magnitude of the radiation dose. In contrast to other physical and chemical methods, at the macroscopic and microscopic level, when the exposure process is properly designed, there is little concern that any portion of the material treated or any portion of the microbial population has not been exposed to the radiation. In part, for this reason, ionizing radiation is used extensively by the medical device manufacturers to sanitize and sterilize their products, some of which have complex geometries [1, 2]. The ionizing radiation used includes high-energy electrons from linear accelerators (LINAC) and gamma photons, generally from ^{60}cobalt and ^{137}cesium sources and in some cases x-rays, including those produced by of LINAC electrons.

The industrial applications and means of radiation for sanitation/sterilization are well-established [1-4] and Papers within this volume. This is in part due to experiments since the 1940s to examine the ability of ionizing radiation to inactivate viruses [7]. The viruses studied have generally been those that pertain to food or waste sanitation and decontamination or used as representatives of broader classes of pathogens [8-21]. Much of this information has been recently summarized [22]. Some foods, notably spices, dairy products and ground meats are now routinely being treated to reduce or eliminate the microbial burden [4]. The most recent and currently prominent use of ionizing radiation has been by the US Postal Service to treat mail contaminated with *B. anthracis* spores.

Despite the extensive industrial use of ionizing radiation there are important potential limitations to the knowledge base as applied to BW/BT neutralization. Not surprisingly, relatively few radiation sensitivity studies are for putative BW/BT agents or even closely related phylogenetic types. Most studies have used ^{60}cobalt gamma photons; there are two reports for neutrons [18, 22] and, to our knowledge, no reports in the primary scientific literature for high-energy electrons. For industrially relevant processes the anticipated bioburden is low, as medical device fabrication or food handling processes are either relatively hostile environments for viruses and/or designed to minimize contamination. Therefore the recommendations for log reduction factors, generally 1×10^6, are likely too low. Concentrations that are achievable are approximately10^7 to 10^9, even without concentration, and infectious doses for some viruses are estimated to be as low as 100 or less. Combined with a safety margin of 2-3 logs implies dose reductions of 10 to 12 logs would be needed. A complicating factor is the potential form of BW/BT agents ranges from concentrated bulk materials to well dispersed aerosols to surface contamination at potentially widely varying concentrations. Further, there appear to be no reports on how additives, including putative radioprotectors or radiosensitization agents alter viral responses to radiation. Finally, information on the effects of physical parameters, such as hydration and exposure temperature, are extremely limited and primarily based on comparing studies using widely different experimental conditions and virus types, making data interpretations problematic [22].

2.2. Radiation Sensitivity - Need for Empiricism

In determining radiation sensitivity, two of the most important parameters are the D_{10}, decimal reduction value and the shape of the inactivation curve. The D_{10} value is the dose necessary to cause a one-log reduction in titer and is a common method of expressing the susceptibility of viral, and other microbial pathogens, to inactivating treatments. Viruses are relatively radiation resistant with most D_{10} values reported between 1 and 10 kGy (kilo-Gray) for ^{60}cobalt gamma photons. Most studies have been conducted with ssRNA (single stranded RNA) viruses but some results are available for other genome structures of both RNA and DNA viruses. Using these and other data for cross comparisons is problematic as the broad range likely reflects not only the true variability in radiation sensitivity, but also the exposure conditions, the viral preparation methods, and the methods of titration used in different laboratories [22]. For some older literature, the radiation dosimetry was performed using very different techniques and parameters. For example dose measurements in the commonly reported Roentgens, (i.e. in terms of exposure) require assumptions about the experimental conditions, which may not be well described, to make conversions into dosimetry units of Gray or Sievert.

Table 1 illustrates these points using reports of the radiation sensitivity of vaccinia virus (VV). VV is the member of the Orthopox family extensively used for research *in vivo* and *in vitro*. *Variola major*, the causative agent of smallpox, is the family member of greatest concern. Studies included are some of the earliest results on ionizing radiation effects on a virus, and the studies span a 30 year period. Each study used different VV strains, preparative methods, sample temperature and hydration conditions and dose rates (not shown). The range of D_{10} values is broad, from about 1 to 8 kGy, and nearly equivalent to the range reported for the majority of viruses (Figure 1). At this point in time it is essentially impossible to determine whether the cause for this variation can be attributed to any one or another experimental difference between the studies.

Unfortunately, there is not a strong basis on which to precisely predict radiation sensitivity of viruses. Based on the idea that viral inactivation is likely primarily due to radiation induced nucleic acid damage, it was hoped by grouping the D_{10} values by virus genomic structure and by exposure conditions that values would cluster. The exposure conditions are important, as hydrated material should be more susceptible, with attack presumptively occurring by both aqueous free radicals and direct radiation damage. Direct damage is the expected mechanism for viruses in the frozen state. As Figure 1 shows, values did not cluster with either parameter. Virus titer methods included both plaque and tissue culture infectious dose (TCID) assays, but for clarity the data has not been separately plotted as no clustering was seen with either titer method.

Radiation target theory would predict that the larger the viral genome is, the lower the D_{10} will be, as the larger genome would incur more damage at a lower dose relative to a smaller viral target. Individual reports have demonstrated rough correlations between titer reduction and radiation dose [13]. And an earlier literature suggested that radiation sensitivity could be analyzed on the basis of the production of sufficient ionization pairs within the target, e.g. nucleic acid volume. Others however, demonstrated correlations between the radiation sensitive volume and actual physical measurements and appeared to be valid only over a limited size range [7]. Chemical kinetic models for heat [23] and radiation and heat inactivation of viruses based on assumptions about intermediate inactivation states have been formulated for viruses with some success [24, 25].

This approach should be further investigated to determine if it could be used to provide better predictive calculations of virus radiation sensitivity than earlier target theory analysis. The data collected of D_{10} values from medically relevant viruses was used in hopes that an empirical relationship between molecular weight (MW) and radiation sensitivity would be

Table 1. Comparison of D_{10} Values for Vaccinia Virus Ionizing Radiation Exposure and Experimental Conditions.

Author Year	Radiation Quality	D_{10}	Exposure Conditions	Virus Strain	Virus Preparation
Decker 1969ab	Gamma	6.04	-70C serum	Merieux 37 (Lister)	Gradient purified
Palacios 1963	Gamma	8.14	RT serum glycerol water peptone	Lancy	Chorio-allontoic dilution
Friesen 1963	Gamma	4.18	RT serum	IHD	
Wilson 1961	Gamma	2.62	RTdry	Levanditi	Ca in PBS lyophilized
McCrea 1960	Gamma		RT dry 0°C		Lyophilized in water
Lea 1942	Alpha	2.62	RT? dry	Lister	Dry in Gelatin On glass
	Gamma	5.25	RT? dry	Rabbit	
	X-ray	2.10	RT? dry	Attenuated	
Gowen 1939	X-ray	3.16-6.3	RT saline		

RT- Room Temperature / Ambient. PBS - Phosphate Buffered Saline. D_{10} value is the principle rate constant re-estimated from the reported data in Roentgens and as D_{37} values. Blank cells indicate parameter could not be determined. See primary references for additional details of exposure conditions. [7, 32-38]

discernable across the currently reported values. However, the aggregate virus data shows a less strong relationship than that reported by individual studies, and the predictive value appears to be weak, as illustrated in Figure 2.

A very important limitation for most reported D_{10} values is that they are initial slope rate constants, having been derived from reduction in titer of only several log cycles not to levels within the limit of sensitivity for the titer assay. The practical issue is that as the D_{10} is a slope constant, even modest changes in its value could alter the level of inactivation actually achieved. Testing to the limit of virus titer assays can be both time consuming and costly. For example, a virus with a D_{10} of 5 kGy and an initial concentration of 10^9 pfu/ml requires approximately 40 kGy of dose to reduce the sample content to 10 pfu/ml, a readily determined limit for virus detection. Even for sources with dose rates on the order of 10

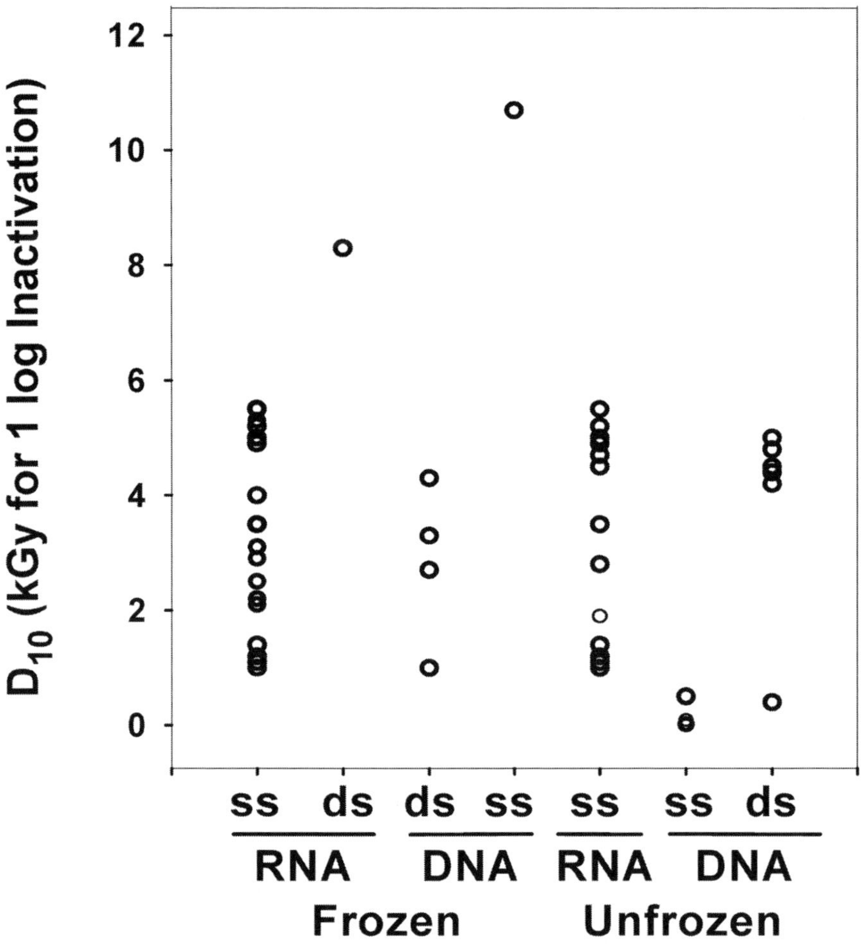

Figure 1. Summary of D_{10} values from published reports. Values have been grouped by virus genome structure and by conditions during radiation exposure. The magnitudes of D_{10} values do not appear to show any correlations relative to these parameters. The overall range is broad, 1 to 14 kGy, but most values are between 1 and 6 kGy. ds – double strand ss – single strand genome structure (From [22]).

kGy per hour, experimental times would be relatively long and potentially costly. This calculation assumes that the inactivation curve is log-linear, e.g. a single exponential describes the inactivation-dose relationship. Similarly, predictions of the dose necessary for titer reduction to the desired safety level using these initial rate D_{10} values assume the dose reduction curve is a single exponential. Presence of either a significant initial shoulder or a long tailing effect at low titer-high radiation doses, e.g. a complex inactivation curve with multiple slope constant values (D_{10}s), could significantly change the actual dose necessary to reduce virus titers to acceptable levels. Most reported curves for viruses do show this single exponential, log-linearity, without a shoulder. An important exception is vaccinia virus, which is reported to have an initial high rate with a more dominant slow component. Tailing effects have not been reported for radiation inactivation of viruses but nor have they been well investigated, likely due to the cost and time of protracted exposures necessary to probe the relevant region of the dose reduction curves.

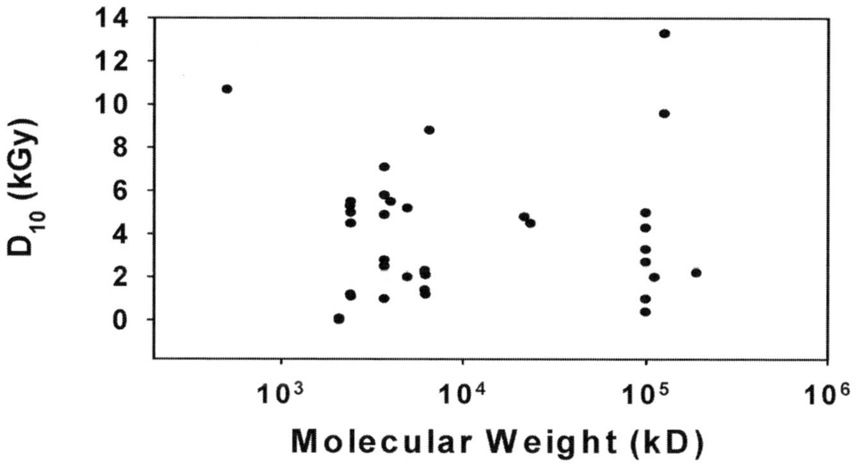

Figure 2. Comparison of D_{10} and Viral Molecular Weight. Data was summarized from the literature and recent experimental work at AFRRI [22]. Genome size in base pairs (bp) was taken from Virology LabFax or the National Library of Medicine Genome database. Molecular weights were calculated in kiloDaltons (kD) using 649 daltons per base pair for double-stranded genomes or 324 daltons per base for single-stranded genomes. Most D_{10} values are within the 1- to 6-kGy range. The correlation between D_{10} value and molecular weight is weak.

Therefore, application of radiation to inactivate viruses can be designed using either a broad approach or a specific one. Current data on radiation sensitivity suggests that most virus D_{10} factors are between 1 and 6 kGy, with a few being higher. The easiest means of designing inactivation methods is to assume any viral contaminant present has a radiation resistance as great as, but not greater than, what has been generally reported. For the majority of virus types this would be 6 kGy, not the highest values, up to14 kGy. These higher values are few, but include important viral types, notably vaccinia virus and HIV. The most important problem with this approach is if this assumption is incorrect, then material could still have significant viral bio-burdens. Alternatively, if the actual contaminant has a lower sensitivity, use of high total doses of radiation is more time consuming and costly and damaging to the materials being decontaminated then necessary for safety. The alternate approach, of course, is to use a specific targeted approach. Assuming that the virus types are known, experiments to determine the D_{10} could be performed. Ideally they would be performed using samples with characteristics as close to those of the material to be decontaminated, using the same radiation source, or a closely matched one.

2.3. Experimental Design Considerations

Factors affecting determination of the efficacy of radiation inactivation can be divided into those primarily associated with the type and source of radiation and those that are primarily virological. Furthermore, both in the laboratory and in real-world applications there can be significant design/engineering issues to solve to accomplish the exposure of samples or contaminated materials. Radiation parameters include the type of radiation, energy spectrum, dose rate and total dose needed. Virological issues include the virus type, methods of determining titer or biochemical activity, the physical state of the virus sample, including the temperature and hydration state, and the presence of other materials, particularly those which could act as radio-protectants or -sensitizers. Potential problems for comparing radiation inactivation values from differing laboratories include the titer method selected, the inherent variability of viral quantitative (plaque forming units) and quantal (cytopathic) assays, and differences in the exposure conditions. The physical experimental setup can greatly alter the radiation dose and quality actually delivered and/or the amount of direct versus secondary free radical damage to the sample. Physical conditions can affect how the exposure occurs, bilateral versus unilateral exposures being a simple example. Safe containment of active viruses within radiation sources, maintaining desired temperature and hydration state, and using radiation compatible materials are all factors complicating accomplishing exposures consistently, accurately and with good dosimetry measurements.

3. Recent Work – Experimental Design Considerations and Results

3.1. Radiation Source

Details for our experimental designs have been published and described here only briefly [22, 26]. The majority of experiments have been conducted using ^{60}cobalt sources, although other important qualities of radiation, high-energy electrons and neutrons have been used on a limited basis. One goal of these studies was to provide cross comparisons of virus sensitivity for different qualities (types) of radiation, while minimizing other parameters. Therefore, data was obtained at varying total doses, keeping the dose rate constant, in so far as possible. The dose rate chosen was the best operational point between the AFRRI source capabilities and the long exposure times necessary. The experiments were conducted at or near the maximum rate available for the cobalt source, 65 Gy/min, as this rate was above the minimum ones for the TRIGA reactor or linear accelerator and minimized the time for high dose exposures. Future dose rate studies that also involve different qualities will need careful planning so that the dose rates chosen neither exceed or are below the different source capabilities. Dose rate and source characteristics should be given careful consideration in future work relative to cross-comparison with existing data, especially if experiments are being conducted to characterize a particular radiation device or real-world exposure situation.

Other important design issues are those common to all radiobiology experiments such as methods for accurate dosimetry, radiation field size and homogeneity in comparison to the biological sample form factor. The sample packaging and form factor for virus exposure requires careful consideration as it can differ depending on the amount of agent, the need for safety packaging, which is governed by the biosafety level of the agent, and the need for maintaining virus viability. Viruses are sensitive to light, heat, and desiccation [27] and therefore these parameters must be controlled during radiation exposure and transport to and from the source facility to avoid confounding inactivating effects on the virus titer. Containers designed to maintain the viruses must also be compatible with the radiation sources. For example, they should be of low radiation cross-section to avoid confounding scattering and Bremstrung effects and many sources have specific dimensional requirements for sample insertion and removal that must be accommodated. Experiments at AFRRI have been conducted using both a custom made device and standard infectious disease secondary safety containers used for shipping microbiological agents and medical samples. The custom device was constructed of Styrofoam® having separate compartments for sample vials and dry ice, [28] Appendix C). Sample cooling was via the vapor phase and maintained temperature for up to 6 h. As the dry ice is in a separate compartment, changes in its volume did not alter the exposure conditions for the samples. Infectious disease secondary containers were tested to very high doses, up to 100 kGy and shown to maintain their integrity. Their thermal conductivity allows samples to be maintained at any desired temperature below ambient by external cooling, e.g. packing in dry ice. In circumstances where extra containment was desired, we have used additional secondary containers or heat sealable cryo-sample tubing around the primary sample vial.

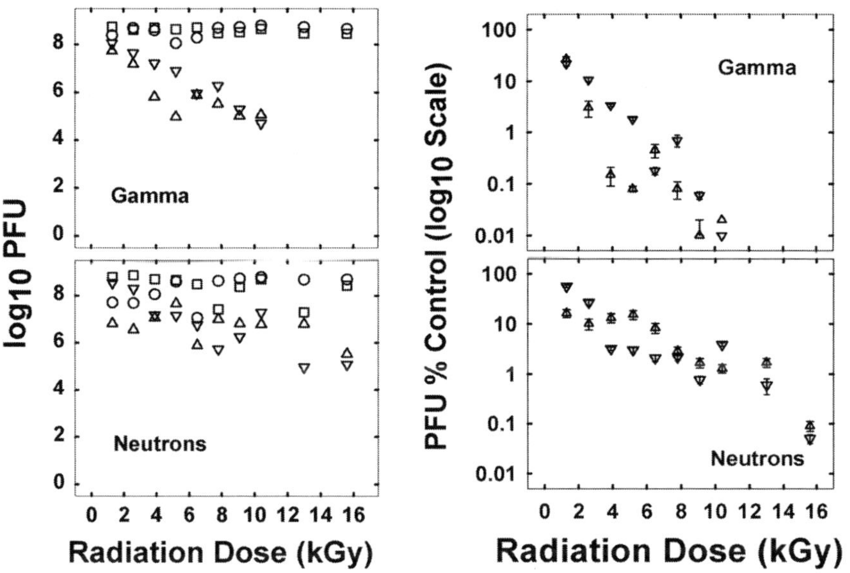

Figure 3. Radiation Titer Inactivation Data for Influenza A Virus. Plaque data (PFU; left panels) is shown for controls for two closely related strains of influenza A virus (circles and squares; X31/PR8) and 60cobalt gamma photon and neutron exposed virus (upright and inverted triangles; X31/PR8). The same data normalized to the dose-paired controls are also shown (PFU % Control; right panels). The PFU data illustrates the inherent variability of these measurements. Note that due to the lower effectiveness of neutrons much higher doses are needed to observe differences between exposed and mock exposed samples. Use of normalized data using dose-matched controls provides a good means for reducing experimental noise levels. (Re-drawn from [22], see for additional details).

3.2. Virology

3.2.1. Influenza Virus

Influenza A virus has been used in most of the recent experiments conducted at AFRRI [22]. Influenza virus is an important pathogen in its own right, resulting in the well-known yearly worldwide pandemics. It is a 100 nm, enveloped virus with a single stranded segmented RNA genome. Many of the putative BT/BW agents have similar characteristics, as they are similarly sized enveloped ssRNA viruses, with both single stranded and segmented genomes. Potentially, inactivation would be more difficult for viruses with segmented genomes, as super-infections, cell infection by more than one virus, could result in a higher probability of rescue of both, even if both had radiation damage, but in different genes. Conversely, single-stranded RNA genomes would be inactivated by a single lesion, in the absence of complimentary strands or repair enzymes, as is the case for higher organisms' DNA repair mechanisms. Recently it has been suggested that influenza could be used as a BT/BW agent [29] and has been included by the United States Department of Health and Human Services as a Group C biological agent (see www.niaid.gov). Importantly, it is a BSL2 virus, making it safe to use in all the source facilities without meeting special regulatory requirements and reducing concerns about safety and decontamination if inadvertent virus sample containment failure occurred.

Figure 3 illustrates an important problem in obtaining accurate D_{10} values. Virus titer assays are inherently noisy as illustrated by the 1 log range of values in the paired mock exposed controls. To determine an estimate of the D_{10}, sufficiently high exposures are

Table 2. Summary of D_{10} Values for Influenza A Virus Combination of Virus Strains, Radiation Quality and Titer Method.

Virus Strain	X31		PR8	
Radiation Quality	**PFU**	**TCID**	**PFU**	**TCID**
^{60}Cobalt Gamma Photons[1]	**2.5±1.5**	**7.1 ± 0.43**	**2.8 ± 0.09**	**5.8 ± 0.20**
Neutrons[1]	**5.5 ± 0.27**	**15.1 ± 1.9**	**4.9 ± 0.54**	**14.2 ± 1.3**
	X31 & PR8 Combined			
High-Energy Electrons[2]	**ND**	**5.6 + 0.2**		

needed to cause a titer reduction that is clearly statistically different from the controls. For a low D_{10}, as illustrated by the gamma photon effects (Fig 3 top left panel), a relatively low total dose can provide good separation between control and experimental titer values from which to estimate the D_{10}. Conversely, the effect of neutrons (Fig 3 bottom left panel) illustrates that for a high D_{10}, titer reductions which clearly differ from the control values require much higher radiation doses and is only evident at approximately 10 kGy. Normalization and pairing techniques as well as applying more sophisticated statistical technique [26] can be useful, but ultimately only sufficiently high doses can provide good estimates.

D_{10} values are also sensitive to the end point analysis used. Common assays are plaques (plaque forming units per ml; pfu/ml) and CPE (cytopathic effect measured as the virus dilution causing an infection in 50% tissue culture host cells e.g. $TCID_{50}$) that are used to determine and compare virus viability exposed to different anti-viral agents. *In vivo* assays, biochemical assays and production of viral antigens by host cells are also used. As all these endpoints involve related but different interactions between the virus and host there is the potential for different estimates of virus radiation sensitivity to be obtained. Table 2 presents the D_{10} values derived for three different types of radiation using both types of assays for the gamma photon and neutron exposures. The D_{10} value is considerably higher using the CPE assay compared to the plaque assay for both qualities of radiation. These differences are large enough to alter the estimate of total dose needed for inactivation. For example, consider the values for gamma photons only, but for both strains and titer methods. Assuming a titer of 10^8 pfu/ml, total doses as low as 20 kGy or as high as 60 kGy would be needed to reach < 1 PFU, which at 10 kGy/h is either 2 or 6 hours of exposure.

The data in Table 2 provides additional insight into virus ionizing radiation sensitivity and its measurement. The advantage of cross comparisons of a single laboratory's data is evident in that values derived from any given radiation quality-titer method. The values are very similar and the error estimates are low and consistent (c.f. Table 2 in [22] and references therein). As expected [5, 6, 30, 31], D_{10} values for gamma photons and high-energy electrons are similar. Unexpectedly, neutrons were much less effective then either of the other radiation qualities. Typically, neutrons have greater damaging effects on biological systems and at present the reason for this "reverse" effectiveness is unclear [22]. However, relative to neutralization of putative viral BW/ BT agents, these data suggest that neutrons are not the best choice.

3.2.2. Vaccinia Virus

Recently, a limited study has been conducted with vaccinia virus [28], in response to the finding of the wide range of reported D_{10} values (Table 1). The strain used was Western Reserve (WR); while neither a vaccine nor highly virulent strain, it is widely used for research on the Orthopox viruses including extensive use in recombinant gene expression systems both *in vivo* and *in vitro*. Purified samples of frozen hydrated virus were exposed to ^{60}cobalt gamma photons over a dose range of 0 to 20 kGy to obtain an initial slope estimate of the D_{10}. The inactivation curve had only a small indication of the two-component curve previously reported by some investigators (see Table 1 References 32, 33). Therefore, a single exponential fit was used; resulting in a D_{10} estimate of 9.6 ± 2.3 kGy. This estimate of VV radiation sensitivity places it at the upper end of reported values (Figure 1). As Orthopox viruses are of such potentially great concern it will be important to validate the applicability of this estimate of radiation sensitivity for neutralization.

4. Conclusions

Various investigators have examined the inactivation of medically relevant viruses by ionizing radiation. The broad outlines of their sensitivity, as reflected by D_{10} values, are between 1 and 10 kGy, placing viruses between most biological organisms, with sensitivities generally less then 1 kGy, and macromolecules and chemicals with sensitivities greater than 100 kGy. Empirical methods are likely the best means of determining radiation sensitivity data necessary to develop a radiation-based method for inactivation of any particular putative BW/BT viral agent. Careful attention to the radiation source conditions is also necessary, as there is little information on classic radiobiology parameters other than total dose.

5. Acknowledgements

This work could not have been conducted without the program support of the Armed Forces Radiobiology Research Institute (AFRRI), especially from Drs. Greg Knudson, Terry Pellmar, and David Ledney. Important contributors to the experimental work include members of Dr. Lowy's laboratory, Dr. David LaBarre, Ms. Michelle Pignone, and Mr. Joshua Dee and members of the AFRRI Radiation Sources Department for dosimetry and exposures. Mr. William Jackson provided aid in statistical analysis. Ms. Michelle Callahan and LT Aaron Moore provided helpful suggestions in the writing and editing of the manuscript. Invaluable collaborators were Dr. Marc F. Desrosiers, National Institute for Standards and Technology and Dr. Christopher C. Broder, Uniformed Services University of the Health Sciences.

References

[1] Block, S.S., editor. 2000. Disinfection, sterilization, and preservation, 5th ed. Lippencott Williams & Wilkins, Philadelphia.
[2] Russell, A. D., Hugo, W. B., and Ayliffe, G. A. J., editors. 1999. Principles and Practice of Disinfection, Preservation and Sterilization, 3rd ed. Blackwell Science, Oxford.
[3] International Consultive Group on Food Irradiation, editor. 1992. Training manual on operation of food irradiation facilities, IGFI Document ed. Vol. 14 1st ed. ICGFI Secretariat Joint FAO/IAEA Divsion of Nuclear Techniques in Food and Agriculture, International Atomic Energy Agency, Vienna, Austria.
[4] Farkas, J. 1998. Irradiation as a method for decontaminating food. Int. J. Food. Micro. 44:189-204.
[5] Barendsen, G. W. 1994. The relationship between RBE and LET for different types of lethal damage in mammalian cells: biophysical and molecular mechanisms. Rad. Res. 139:257-270.
[6] Ward, J. F. 1988. DNA damage produced by ionizing radiation in mammalian cells: identities, mechanism of formation, and reparability. Progress in nucleic acid research and molecular biology 35:96-125.
[7] McCrea, J. F. 1960. Ionizing radiation and its effects on animal viruses. Annal. N.Y. Acad. Sci. 83:692-705.
[8] Daley, J. P., Danner, D. J., Weppner, D. J., and Plavisic, M. Z. 1998. Virus inactivation by gamma irradiation of fetal bovine serum. Focus (LifeTechology) 20(3):86-88.

[9] Gamble, W. C., Chappell, W. A., and George, E. H. 1980. Inactivation of rabies diagnostic reagents by gamma radiation. J.Clin. Micro. 12(5):676-678.
[10] Gruber, J. 1971. Immunogenicity of purified venezuelan equine encephalitis virus inactivated by ionizing radiation. Infect. Immun. 3(4):574-579.
[11] House, C., House, J. A., and Yedloutschnig, R. J. 1990. Inactivation of viral agents in bovine serum by gamma irradiation. Can. J. Micro. 36:737-740.
[12] Sullivan, R., Fassolitis, A. C., Larkin, E. P., Read Jr, R. B., and Peeler, J. T. 1971. Inactivation of thirty viruses by gamma radiation. Appl. Micro. 22(1):61-65.
[13] Thomas, F. C., Davies, A. G., Dulac, G. C., Willis, N. G., Papp-Vid, G., and Girard, A. 1981. Gamma ray inactivation of some animal viruses. Can. J. Micro. 45:397-399.
[14] Thomas, F. C., Ouwerkerek, T., and McKercher, P. 1982. Inactivation by gamma irradiation of animal viruses in simulated laboratory effluent. Appl. Environ. Micro. 43(5):1051-1056.
[15] White, L. A., Freeman, C. Y., Hall, H. E., and Forrester, B. D. 1990. Inactivation and stability of viral diagnostic reagents treated by gamma radiation. Biologics 18(4):271-280.
[16] Megumi, T., Fujita, S.-I., Iwai, Y., and Ito, T. 1993. The effect of gamma-ray-induced radicals on activities and membrane structure of sendai virus in aqueous solutions. Rad. Res. 134:129-133.
[17] Rosen, A., Taylor, D. M., and Darai, G. 1987. Influencing of ionizing radiation on herpes simplex virus and its genome. Int. J. Radiat. Biol. 52(5):795-804.
[18] Singh, S. P., Cohen, D., Dytlewski, N., Houldsworth, J., and Lavin, M. F. 1990. Neutron and gamma-irradiation of bateriophage M13 DNA: use of standard neutron irradiation facility (SNIF). J. Radiat. Res. 31:340-353.
[19] Toyoshima, K., Niwa, O., Yususdo, M., Sugiyama, H., Tahara, S., and Sugahara, T. 1980. Sensitivity to gamma rays of avain sarcoma and murine leukemia viruses. Virology. 105:508.
[20] Elliot, L. H., McKormick, J. B., and Johnson, K. M. 1992. Inactivation of lassa, marburg, and ebola viruses by gamma irradiation. J.Clin. Micro. 16(4):704-708.
[21] Lupton, H. W. 1981. Inactivation of Ebola virus with 60Cobalt irradiation. The Jounal of Infectious Diseases 143(2):291.
[22] Lowy, R. J., Vavrina, G. A., and LaBarre, D. D. 2001. Comparision of gamma and neutron radiation inactivation of influenza A virus. Antivir. Res 52:261-273.
[23] Woese, C. 1960. Thermal inactivation of animal viruses. Annal. N.Y. Acad. Sci. 83:728-741.
[24] Trujillo, R., and Dugan, V. L. 1972. Synergistic inactivation of viruses by heat and ionizing radiation. Biophysical J. 12:92-113.
[25] Petin, V. G., and Komarov, V. P. 1997. Mathematical description of synergistic interaction of hyperthermia and ionizing radiation. Math. Biosci. 146:115-130.
[26] LaBarre, D. D., and Lowy, R. J. 2001. Improvements in methods for calculating virus titer estimates from TCID50 and plaque assays. Journal of Virological Methods 96:107-126.
[27] White, D. O., and Fenner, F. J., eds. 1994. Medical Virology, 4th ed. Academic Press, New York, NY.
[28] Lowy, R. J., Broder, C. C., Feng, Y.-R., Desrosiers, M. F., and Elliott, T. B. 2003. Ionizing radiation of vaccinia virus using gamma photons. AFRRI SP 02-03:1-40.
[29] Krug, R. M. 2003. The potential use of influenza virus as an agent for bioterrorism. Antivir. Res 57(1-2):147-150.
[30] Johns, H. E., and Cunningham, J. R., editors. 1983. The physics of radiology. Charles C. Thomas, Springfield, IL.
[31] Hall, E. J. 1994. Linear energy transfer and relative biological effectiveness. In Radiobiology for the radiologist, 4th ed. J.B. Lippencott, Philadelphia. 153-164.
[32] Palacios, R., Contreras, G., Espejo, R., Jimenez, R., Ohlbaum, A., and Toha, J. 1963. Compound survival curve of vaccinia-virus after gamma radiation. Biochimica et Biophysica Acta:149-151.
[33] Decker, C., Guir, J., and Kern, A. 1969. Infectivity and capacity for DNA replication of vaccinia virus irradiated by gamma rays. J. Gen. Virol. 4:221-227.
[34] Decker, C., Guir, J., and Kirn, A. 1969. Dissociation by gamma rays of vaccinia virus functions. Rad. Res. 40:520-524.
[35] Friesen, J. D., Sankoff, D., and Lminovitch, L. 1963. Radiobiolgy studies of vaccinia virus. Virology. 21:411-424.
[36] Wilson, D. E. 1961. Radiation inactivation of vaccinia virus. Rad. Res. 14:796-802.
[37] Lea, D. E., and Salaman, M. H. 1942. The inactivation of vaccinia virus by radiation. Brit. J. Exp. Path. 23:27-37.
[38] Gowen, J. W., and Lucas, A. M. 1939. Reaction of variola vaccine virus to roentgen rays. Science 90(2348):621-622.

Detection and Prevention of Bioterrorism Agents – Portuguese Case Studies

Maria Luísa BOTELHO[a], Sandra CABO VERDE[a], Paula MATOS[a], Pet MAZARELO[b] and Rogério TENREIRO[c]
[a]*Instituto Tecnológico e Nuclear, Estrada Nacional 10, Apartado 21, 2686-953 Sacavém, Portugal*
[b]*Laboratório Militar de Produtos Químicos e Farmacêuticos, Av. Dr. Alfredo Bensaúde, 1899 Lisboa, Portugal*
[c]*Departamento de Biologia Vegetal e Centro de Genética e Biologia Molecular, Faculdade de Ciências da Universidade de Lisboa, 1749-016, Portugal*

Abstract. The Radiation Technology Group of ITN Portugal is developing two different projects that could contribute to the prevention and early detection of microbiological war: "Control of the environment in surgical rooms at Army hospitals and the impact on the incidence of cross-infection" and "Sanitation of chicken eggs by ionising radiation".

The first project focuses on the development and improvement of alternative techniques to control the environment in surgical rooms leading to the detection and identification of nosocomial microorganisms in a Portuguese Hospital. A database can then be constructed that could demonstrate the relation between the improvement of the airborne conditions and the hospital infection agents. This project includes a continuous monitorization of the surgery rooms' natural air bioburden and nosocomial microorganisms, and the consequent updating of the database. This study will be carried out based on the molecular type of isolated strains from infected patients during surgery and the correlation with surgical environment isolated strains. This procedure could lead to a model for the detection of emerging microorganisms that could become hospital infectious agents.

The second project uses the decontaminating capacity of ionising radiation to inactivate pathogenic microorganisms in eggs and will be presented in a different section.

These two studies, although in different microbiological areas, are examples of how we could deal with potential harmful public health agents: detection and prevention.

Introduction

Bioterrorism is defined as the use of biological agents to inflict disease and/or death directly on humans, or in their surroundings so as to affect human life, such as the use of animals or plants as vectors. This threat has existed for along time, even before Leeuwenhoek (1752-1833) and Pasteur (1822-1895) introduced knowledge about microbes and the detection and prevention of infectious diseases. Historical evidence suggests that Greeks, Romans and Persians attempted to pollute their enemies drinking water supplies [1].

Nowadays, the emerging of infectious diseases could be caused by outbreaks of food-borne and water-borne infections, hospital-acquired (nosocomial) infections and antibiotic and/or disinfectant resistance as well as biological weapons [2].

There are several specimens described as suspected of containing selected biological terrorism agents [*in http//www.*bt.cdc.gov/bioagents.asp]. There are recommended protocols for the detection of these agents. Nevertheless, we have to be aware and develop programmes of surveillance that effectively lead to the detection and prevention of any outbreaks on populations, especially those who are exposed to the threat of the war.

Thus, the detection and the prevention of biological agents that can cause an endemic or epidemic must continue as well as the development of protocols and programmes to detect the incidence of an outbreak as early as possible.

There are a variety of motives for these kinds of actions in war, revenge and/or political destabilization. They may target the civilian population to create panic and threaten civil order. An example is the response to the scare of envelopes containing *Bacillus anthracis* mailed in the USA, despite the actually limited dissemination [1]. Despite the fact there were few cases of specific symptoms of the illness, the scare led to disruption and public anxiety. Fear and anxiety may contribute to the reduction of natural protection to human health.

Some authors [1] consider the most efficient method of delivering biological agents the air-borne route: dispersing agents in aerosols and contaminated food. Although airborne attack is subject to major uncertainties, such as air movements, conditions in enclosed spaces and other factors. Thus bioterrorism is uncontrollable and puts at risk the entire population, including the people that released the hazard.

Biocontamination controlled by air sampling could lead to the reliable detection and measure of the impact, enabling corrective actions to improve the surroundings. Therefore environment control based on the Hazard Analysis Critical Control Points (HACCP) is a target to pursue to get a better rastreability and correct actions for the improvement of protocols [3]. This method has been widely recognized as the preferable system ensuring a better control in hospital infections and food safety. HACCP is based on three principles: identification and assessment of hazards, determination of the critical control points (CCP) to control any identified hazards and the establishment of systems to monitor the critical control points [4], [5].

The Radiation Technology Group of ITN Portugal is developing two different projects that were designed irrespectively of terrorism but could contribute to the prevention and the early detection of microbiological war: These are: 1) "Environmental control in surgical rooms at Army hospitals and the impact on the incidence of cross-infections" and 2) "Sanitation of chicken eggs by ionizing radiation".

The first project focuses on the development and improvement of alternative techniques to control the environment in surgical rooms leading to the detection and identification of nosocomial microorganisms in a Portuguese Hospital [6]. A database is then established that could demonstrate the relation between the improvement of airborne conditions and infectious hospital agents. This project includes a continuous monitorization of the surgery rooms' natural air bioburden and nosocomial microorganisms, and the consequent updating of the database. This study will be carried out based on the molecular type of isolated strains from infected patients during surgery and the correlation with isolated strains from the surgical environment. This procedure could lead to a model for the detection of emerging microorganisms that could become hospital infectious agents.

The second project applies the decontaminating capacity of ionising radiation to inactivate pathogenic microorganisms in eggs. *Salmonella* spp. and *Campylobacter* spp. are natural contaminants of eggs and the leading causes of bacterial gastroenteritis in humans. In this study, the D_{values} of reference strains of *Salmonella* and *Campylobacter* were determined and sub-lethal gamma radiation doses were applied to artificially contaminated eggs, in order to predict which irradiation dose could guarantee egg sanitation. It was also determined the gamma and electron beam inactivation response of egg bioburden [7], [8], [9]. This project is described in more detail further on in a different presentation.

Although the "Environmental control in surgical rooms at Army hospitals and its impact on the incidence of cross-infections" project started at the end of last year, some results are presented here. In this first phase we are considering the baseline data in order to have the number of microbes, as well as the potential critical points and the resistance of microbiota to the antibiotics prescribed to patients and the disinfectants used in the hygienic protocols. The purpose is to implement improvements to Protocols based on the results, namely the resistance of the nosocomial microorganisms to disinfectants and antibiotics [10].

The use of computer-based networks for the comparison of microorganisms in the future could enable the fast recognition of strains with identical DNA fingerprints. This could lead to the better detection of exposure to common sources of microorganisms and allow the rapid recognition of any connection in sources of contamination [1].

1. Material and Methods

1.1. Sampling

Studies were taken in an operating room theatre before, during and after an orthopedic surgery at an Army Hospital.

The selection of equipment to use in air sampling was previously validated, comparing the efficiency of recovery of the air sampler MAS100 (Merck Air Sampler) against two equivalent equipments SAS (Surfair Air System) and URSA (equipment developed in ITN) for in and outdoors sampling. Air was collected in the surgery room, using the air sampler MAS100 in ten locals after and before surgery, and in just one local (to reduce external interferences), with three replicas, during surgery. The volume of air collected was selected, taking into account previously results of biocontamination obtained by sedimentation plates. A volume of 0.1 m^3 of air was then collected in each local.

In parallel, sedimentation plates during 10 min were settled, in eight locals on the surgery room floor.

1.2. Bioburden Determination and HACCP Studies

Sedimentation plates were placed in parallel sites (n=10). Air is projected directly on to a TSA (Triptic Soya Agar-Merck) plate. Plates were incubated for 14 days at 30 ($\pm$ 2)°C. Microorganisms were counted after 24h, 2 days, 3 days, 7 and 14 days. The microbiota was isolated and classified in eleven groups taking into account the gram stain, oxidase test, catalase and were preserved at -80°C for further identification when necessary.

1.3. Antibiograms

Antibiograms were performed with 30 μg of Tazobac®, Tienam®, Ceriax® and Cefoxitin® hospital antibiotics, administrated to patients. The inactivation response of microorganisms to hospital detergents and disinfectants is being carried out.

2. Results and Discussion

2.1. Bioburden Determination

The biocolector MAS100 aspirate the air through a perforated dish at a caudal of 100 l/min. The microorganism's velocity of impact into the agar surface is approximately 11 m/s; this velocity guaranty that all particles with a size superior to 1 μm are collected.

Sampling was planning previously taking into account critical points as the movement of personal and the air flux, with the purpose to obtain a significant sampling of all room area. In this hospital the insufflate air environment is not controlled. The choice of an orthopedic surgery was taken, due to the complexity of the intervention and the dirtiness associated. The parallel use of sedimentation technique was carried out, since this procedure is monthly performed as air monitoring system at the surgeries rooms of the Army Hospital and could be related with the results obtained in the present sampling.

The number of collected microorganisms is presented in colony forming units per m^3 of aspirated air (cfu/m^3). The number of colony counts on agar was corrected using a specific equipment table, which is based on the principle that if the microorganism's quantity is too high, higher is the probability of several microorganisms fall in the same hole of the collector. The number of colony forming units is then divided for the volume of air collected.

The cfu/m^3 for air collection and cfu/plate in each phase, before, during and after surgery, were calculated taking in account the medium value of all sampling collected in each local. The air bioburden average values (P= 95%) obtained with the biocolector MAS100 were 1900 ± 1000 cfu/m^3, 1100 ± 24 cfu/m^3 and 2300 ± 420 cfu/m^3, for before, during and after surgery, respectively. The sedimentation results were 25 ± 8 cfu/plate before surgery, 22 ± 4 cfu/plate during surgery and 32 ± 5 cfu/plate after surgery. A better visualization of these results is presented in Figures 1 and 2.

The graphics presented above show the same profile for both methodologies (biocolector and sedimentation): the bioburden decreases during surgery and there is an increase after surgery. This fact could be due to a lesser moving during the surgery than before and after the surgery caused by preparation and room cleaning respectively, which involves more dislocation of personnel and doors.

Comparing the bioburden values between the two techniques, efficiency of microorganism's collection is 100 fold higher for MAS 100 than for sedimentation. This result indicates that the air monitoring method applied in the surgery rooms by the hospital (sedimentation) is not the most appropriate to obtain a significant sampling of all natural air borne

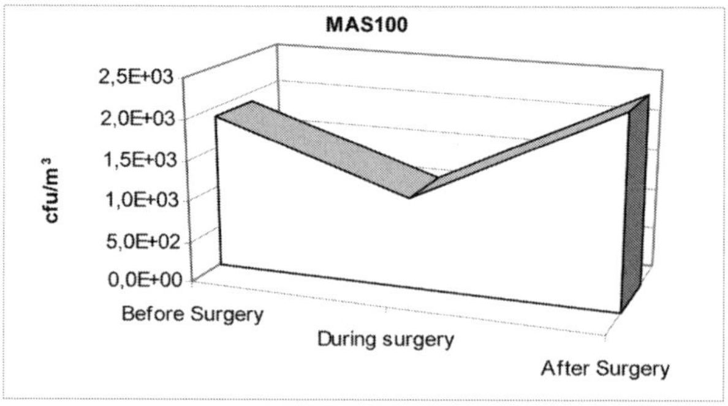

Figure 1. Average values of cfu/m^3 of air bioburden before, during and after an orthopedic surgery in the Army Hospital.

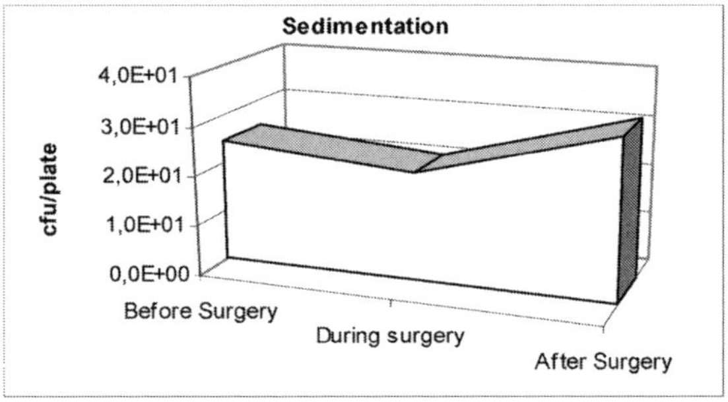

Figure 2. Average values of cfu/plate of air sedimentation before, during and after an orthopedic surgery in the Army Hospital.

microorganisms. That could not lead to a good detection and identification of nosocomial microorganisms.

After morphological characterization of the obtained colonies, it was observed that before, during and after surgery, the major morphological type (50%) was the gram-positive catalase-positive cocci, in air biocolection as well as sedimentation. The other morphological groups identified were gram-negative catalase-positive cocci (18%), gram-negative oxidase-negative rods (15%), gram-positive catalase-positive rods (12%) and gram-negative oxidase-positive rods (3%). This last morphological type was only recovered by means of the biocolector MAS100.

2.2. Antibiograms

In other to evaluate the resistance/sensibility of the recovered strains, antibiograms were performed using four antibiotics administrated to surgery patients. 90% of all strains were susceptible to all antibiotics. One of the strains isolated during surgery by the biocolector, identified as *Pseudomonas fluorescens*, showed resistance to Ceriax® and Cefoxitin® with halos of 0 mm, but susceptibility to the other two antibiotics.

The resistance/sensibility of the environmental strains recovered from antiseptic hospital detergents are being carried out, in order to find out the adequacy of the cleaning protocols to eliminate potential nosocomial microorganisms.

3. Conclusions

The hospital data obtained so far point to the usefulness of the implemented HACCP study to assess infection risk and nosocomial antibiotic and disinfectant resistant strains. Nevertheless, more data should be gathered before we can implement new actions that can effectively reduce the risk of infection. Once the detection of an outbreak can easily be detected, corrective actions can be implemented.

Despite being in different microbiological areas, the "Environmental control in surgical rooms at Army hospitals and its impact on the incidence of cross-infections" and the "Sanitation of chicken eggs by ionizing radiation" studies are two real examples of how we could deal with the detection and prevention of potential harmful public health agents.

As in all health and safety considerations, prevention is the most desirable option. The use of databases in computer networks could lead to faster detection of an endemic or epidemic in a surveillance system that is sensitive for identifying clusters of illness, especially in critical sites, such as the military hospitals. Such systems could permit the identification of disease outbreaks whether these are intentional or unintentional, allowing a solution to be found.

The authors agree with the statement of Klietmannand and Ruoff [1] "...resources would be better spent on mechanisms that would improve the standard of living and health care for all. These efforts would lead to a healthier population with increased resistance to infection and would also diminish the causes of terrorism...". This could be the way towards world peace.

References

[1] Klietmann, W.F. and Ruoff, K. L., 2001, Bioterrorism implications for the clinical microbiologist. Clin. Microbiol. Rev. 14 (2), 364-381.

[2] Abigail A. Salyers and Dixie D. White, 2002, The uneasy truce: never underestimate the power of bacteria. In: Bacterial Pathogenesis – a molecular approach, 2nd Ed. ASM (American Society for Microbiology) Press. Washington D.C. USA.

[3] WHO, 2002, Terrorist threats to food: guidance for establishing and strengthening prevention and response system. In: Food Safety Issues. WHO, Food Safety Department, Geneva, Switzerland, pp. 4.

[4] WHO, 2002, Global strategy for food safety: safer food for better health. In: Food safety issues. WHO, Geneva, Switzerland, pp. 5-8.

[5] Codex Alimentarius Commission, 1995, Guidelines for the application of the Hazard Analysis Critical Control Point (HACCP) system (CAC/GL 18-1993). In: Codex Alimentarius, vol 1B, General Requirements (food hygiene), FAO/WHO, Rome, pp. 21-30.

[6] Catamo, G., Bonaventura G di, Lattanzio, F.M.Piccolomini R., 1999, Monitoraggio microbiologico di aria e superfici in ambiente di prime cure chirurgiche di ambulatorio INAIL In: Giornale Italiano di Microbiologia Medica Odontoiatrica e clinica Vol.. III N° 2 pp. 128 – 134.

[7] Torok, T., Tauxe, R. V., Wise, R. P., Livengood J. R., Sokolow, R., Mauvais, S., Birkness, K. A., Skeels, M. R., Horan, J. M., Foster, L. R., 1997, A large community outbreak of salmonellosis caused by intentional contamination of restaurant salad bars. J. Am. Med. Assoc. 278, 389-395.

[8] WHO, 2002, Risk assessments of Salmonella in eggs and broiler chickens. In: Microbiological Risk Assessment Series 1. WHO, Geneva, Switzerland, pp. 4.

[9] Food and Drug Administration, 2000, Irradiation in the production, processing and handling of food. In: FDA Federal Register, vol 65, number 141, FDA, USA, pp. 45280 – 45282.

[10] Hecker-Wc Ring-Mrozik Limmer-s, 1994, Mortality and Postoperative cause of death in pediatric surgery Langenbecks – Arch-Chir 379 (3), 172-177. CDC, 2003, Bioterrorism agents/diseases *in* http//www.bt.cdc.gov/bioagents.asp.

Foodborne Agents and Bioterrorism Prevention – A Portuguese Case Study on Ionizing Irradiation

Sandra CABO VERDE[a], Rogério TENREIRO[b] and Maria Luísa BOTELHO[a]
[a]*Instituto Tecnológico e Nuclear, Estrada Nacional 10, Apartado 21, 2686-953 Sacavém, Portugal*
[b]*Departamento de Biologia Vegetal e Centro de Genética e Biologia Molecular, Faculdade de Ciências da Universidade de Lisboa, 1749-016, Portugal*

Abstract. Bioterrorism is an event in a civil setting that is equivalent to an epidemic in a medical scenario. The collaboration between medicine and public health is also important in the clinical diagnostics and epidemiological surveillance in response to the emergence of a biological agent.

The malicious contamination of food for terrorist purposes is a real and current threat, and deliberate contamination of food at one location could have global health implications. The key to prevent food terrorism is to enhance existing food safety programmes in order to protect food production systems. Typical food safety management programmes within the food industry include good manufacturing practice and "hazard analysis and critical control point" – HACCP. Food irradiation could be one additional food safety tool that serves as a complement to other food safety technologies such as the HACCP.

With the purpose to apply the food irradiation as a food safety tool, the Radiation Technology group at the ITN is developing a project entitled "Sanitation of chicken egg by ionizing radiation", which applies the decontaminating capacity of the ionising radiation to inactivate pathogenic microorganisms in eggs. *Salmonella* spp. and *Campylobacter* spp. are eggs natural contaminants and the leading causes of bacterial gastroenteritis in humans. In this study, the D_{values} of reference strains of *Salmonella* and *Campylobacter* were determined and sub-lethal gamma radiation doses were applied to artificially contaminated eggs, in order to predict which irradiation dose could guarantee egg sanitation. Based on the results obtained and to guarantee organoleptic acceptable and safe eggs, a radicidation dose of 1.5 kGy is thus proposed.

Monitoring programmes with a rapid follow-up are important if variation in product is seen that could indicate deliberate contamination. Since a reliable detection step to identify the hazard is crucial in any production process, we validated a PCR method to detect *Salmonella* and *Campylobacter* contamination, directly from eggs and egg-products.

Introduction

Threats from terrorists, criminals and other anti-social groups who target the safety of the food supply are already a reality. Food terrorism is defined as an act or threat of deliberate contamination of food for human consumption with chemical, biological or radionuclear agents for the purpose of causing injury or death to civilian population and/or disrupting social economic or political stability. The biological agents referred to are communicably infectious or non-infectious pathogenic microorganisms, including viruses, bacteria and parasites [1]. Deliberate contamination of food by chemical, biological or radionuclear agents can occur at any vulnerable point along the food chain, from farm to table, depending on both the food and agent. For example, in 1984, members of a religious cult contaminated salad bars in Dalles (USA) with *Salmonella typhimurium*, causing 751 cases of salmonellosis. The attack appeared to be an attempt to sicken citizens and prevent them from voting in an upcoming election [2]. Contamination of food in one country can also have a significant effect on health in other parts in the world. In 1989, staphylococcal food poisoning in the USA was associated with eating mushrooms that had been canned in China [3]. Thus, deliberate sabotage of food could have serious economic and trade repercussions. Industries in many sectors could be put out of business, and countries could experience severe economic and trade disruption. In less developed countries, the economic consequences of a terrorist act on food could also affect development and exacerbate poverty as well as food availability. Protecting food production system is thus a current issue, including development and implementation of better food processing technologies, in order to reduce the likelihood and impact of a terrorist attack. Food safety programmes focus increasingly on the farm-to-table approach as an effective means of reducing hazards. This holistic approach to the control of food-related risks involves the consideration of every step in the chain, from raw material to food consumption. Hazards can enter the food chain on the farm and can continue to be introduced or exacerbated at any point in the chain until the food reaches the consumer [4].

Eggs are an important and nutritionally complete food product, easy to obtain at low costs. Pathogens such as *Salmonella* and *Campylobacter* account for over 90% of all reported cases of bacteria-related food poisoning worldwide, and poultry and poultry products are involved in the majority of traceable foodborne illnesses caused by these bacteria [5]. 77% to 80% of *S. enteritidis* outbreaks have been associated with grade A shell eggs, or egg-containing foods [6]. Relatively to thermophilic *Campylobacter* spp. they are considered the leading cause of enteric illness in most developed countries [7].

The Hazard Analysis Critical Control Points (HACCP) has been widely recognized as the preferable system assuring food safety and is based on three principles: identification and assessment of hazards associated from growing to consumption, determination of the critical control points (CCP) to control any identified hazards and establishment of systems to monitor the critical control points [8]. Food irradiation is an additional food safety tool that serves as a complement to other food safety technologies such as HACCP. Previous research has shown that ionising energy at medium doses can eliminate non-spore-forming pathogens such as *Salmonella* in food products [9]. Irradiation could cause very little increase in food temperature during application; thus it is termed “cold processing”. These features make the process of irradiation more attractive for eliminating pathogens in heat-sensitive products like eggs. Previous studies on microbial inactivation and functional and chemical egg properties demonstrated that the process of irradiation can eliminate pathogens from eggs without adversely affecting egg quality [10]. Since 2000, the Food and Drug Administration (FDA) approved the use of up to 3 kGy ionising radiation dose to reduce the level of *Salmonella* in shell eggs [11].

In order to minimize the risk of infection for consumers, microbiological control of the food chain is being increasingly applied. Thus, the availability of reliable, rapid, and internationally accepted test systems for determination of the presence or absence of food-borne pathogens has become increasingly important for the agricultural and food industry, as well as for legislative of food safety. Conventional cultural methods for detecting *Salmonella* and *Campylobacter* spp. involve enrichment in selective broth, followed by isolation on selective differential agar. *Campylobacter* spp. have demanding growth requirements because they need to be incubated under microaerobic conditions (5% O_2, 10% CO_2, and 85% N_2), which makes the task of isolation laborious and costly. Both a primary and secondary enrichment culture are necessary for isolating *Salmonella* spp. from foods. Isolation can therefore be labour-intensive and expensive when large numbers of samples must be processed. Hence, there is need for a sensitive non-cultural detection method for these foodborne pathogens. The application of culture-independent detection methods such as Polymerase Chain Reaction (PCR) may help to overcome the aforementioned problems. In addition, PCR in general provides faster results than conventional culture and has the potential for automation in which many samples are examined in a short period of time [12]. The selected target gene for the detection of *Salmonella* spp. in eggs is the *inv*A gene located on the *Salmonella* pathogenicity island 1 (SPI1). SPI1 is highly conserved in *Salmonella* spp. and encodes a type III secretion system that exports proteins in response to bacterial contact with epithelial cells. The primer-set amplifies a 284 bp sequence of the *inv*A gene [13]. For *Campylobacter* spp., the target gene for detection in eggs is the 16S gene. This gene is highly conserved but is possible to identify regions with some sequence variability among groups of *Campylobacter* spp. The primer-set amplifies a 287 bp sequence of the 16S gene [14, 15].

The aims of this project were to study the food irradiation technology, which could be applied in the final step of egg or egg-products production system as a preventive tool; and the microbial detection power of the molecular methods such as PCR to identify the principal pathogenic egg natural contaminants to be applied as a monitoring program to identify the hazards in the eggs production line. Firstly, it was determined which dose of gamma radiation could guarantee egg sanitation without severe affects on eggs organoleptic properties, applying sub-lethal doses to irradiate eggs artificially contaminated with reference strain of *Salmonella* spp. and *Campylobacter* spp. to find out which minimal dose leads to egg radicidation. For the detection purpose, we validated a PCR method to detect *Salmonella* and *Campylobacter* contamination, directly from eggs and egg-products.

1. Material and Methods

1.1. Sampling

The eggs and environmental samples were randomly collected at four times spaced by six months, from a commercial egg-production farm in Portugal, consisting of environmentally controlled windowless pavilions. Pavilions were equipped with manure and egg collection belts running under and beside the cages, respectively. Belts are connected between the pavilions and they transport the eggs from each pavilion to a central building where grading and packing of the eggs is carried out. For the determination of total bioburden and characterization of natural contaminants, three samples of fifty eggs, sub-divided in ten eggs from five chicken age bands (40-50, 50-60, 60-70, 70-80 and 80-90 weeks), were analyzed. The detection of critical points takes into account water, feed, faeces and swabs from cages and belts from different chicken age pavilions. D_{values} of reference strains and total natural contaminants were determined using 48 and 12 eggs, respectively.

The egg-products samples (liquid-yolk and white) were collected from a Portuguese Industry. For the inactivation studies in the egg-products, the sampling enclosed 55 kg of liquid-white non-pasteurized, and the same quantity of liquid-yolk equally non-pasteurized.

1.2. Bioburden Determination and HACCP Studies

A previously validated method was used: each egg was washed in 200 ml of saline solution with 0.1% Tween 80 (Difco, USA), and stirred by a stomacher (Stomacher 3500, Seward, UK) for 20 min. Five aliquots were plated in Tryptic Soy Agar (Oxoid, UK) and incubated at 30°C. Colony forming units per egg (cfu/egg) were counted for 1 week. Detection of *Salmonella* and *Campylobacter* was performed according to ISO 6579:1993 and ISO 10272:1995(E), respectively.

The different types of colonies obtained in the bioburden determination (before and after irradiation) and in the HACCP studies were macroscopically, microscopically and biochemically characterized, using catalase and oxidase tests, triple sugar agar (Merck, Germany), urease media (Merck, Germany), and in some cases biochemical strips (API system, bioMérieux, France).

1.3. Irradiation Process

Eggs and egg-products (250 ml) were placed in sterile plastic bags arranged in a wooden support (24 x 12 x 7 cm), and irradiated in a cobalt-60 facility with 47 kCi activity (September 2002) at room temperature, at a dose rate of 1 $kGyh^{-1}$. The range of applied doses was 0.5 up to 5 kGy, with a maximum range dose variation of $\pm$ 5% in the overall egg. The absorbed doses were monitored with routine dosimeters (Amber Perspex and Gammachrome YR).

1.4. Inactivation Studies

Eggs were artificially contaminated, externally and internally, with reference strains of *Salmonella typhimurium* (ATCC 14028), *Salmonella enteritidis* (ATCC 13076), *Campylobacter coli* (DSMZ 4689^T) and *Campylobacter jejuni* (DSMZ 4688^T) using *ca.* 10^7 - 10^8 cfu/egg. Egg-products (liquid-yolk and white separately) were artificially contaminated with *Salmonella enteritidis* (ATCC 13076) and *Salmonella arizonae* (isolated from egg-products) using *ca.* 10^7 - 10^8 cfu/ml of egg-product. The range of irradiation doses for determination of the D_{values} was 0.2 up to 1 kGy for *Salmonella* and 0.2 up to 0.7 kGy for *Campylobacter*. The 90% inactivation of natural contaminants of egg and egg-products was determined with a range of irradiation doses of 0.5 up to 5 kGy in the gamma facility. The number of survivors and the initial contamination (0 kGy) were calculated as described for bioburden determination, using three replicates for each dose.

1.5. Detection of Salmonella spp. and Campylobacter spp. by PCR Directly from Eggs and Egg-products (adapted from Food PCR Project QLRT-1999-0026 protocols [16, 17])

1.5.1. Preparation of DNA Samples

A 25 ml sample of eggs or egg-products artificially contaminated with *S. enteritidis* (ATCC 13076) or *C. jejuni* (DSMZ 4688^T) using *ca.* 10^7 - 10^8 cfu/egg was enriched in:

- 225 ml of Peptone Water (Merck, Germany) and incubated 18 h at 37°C for *Salmonella.*
- 225 ml of Campylobacter Enrichment Broth (Oxoid, England) and incubated 18h at 42°C in microaerobic atmosphere (CampyGen, Oxoid, England) for *Campylobacter.*

A 1-ml aliquot of the enriched culture was centrifuged at 10,000 *g* in a microcentrifuge tube for 5 min. The supernatant was carefully discarded, and cell pellet was suspended in 300 µl of TE buffer (10 mM Tris-HCl, 0.1 mM EDTA [pH 8.0]). The microcentrifuge tube was incubated for 10 min at 100°C in a water bath and immediately chilled on ice. After centrifugation at 14,000 *g* at 4°C for 5 min, the supernatant containing DNA was carefully transferred to a new microcentrifuge tube. A 5-µl aliquot was used as a template DNA for the PCR.

References

[1] WHO, 2002, Terrorist threats to food: guidance for establishing and strengthening prevention and response system. In: Food Safety Issues. WHO, Food Safety Department, Geneva, Switzerland, pp. 4.

[2] Torok, T., Tauxe, R. V., Wise, R. P., Livengood J. R., Sokolow, R., Mauvais, S., Birkness, K. A., Skeels, M. R., Horan, J. M., Foster, L. R., 1997, A large community outbreak of salmonellosis caused by intentional contamination of restaurant salad bars. J. Am. Med. Assoc. 278, 389-395.

[3] Levine, W. C., Bennet, R. W., Choy, Y., Henning, K. J., Rager, J. R., Hendricks, K. A., Hopkins, D. P., Gunn, R. A., Griffin, P. M., 1996, Staphylococcal food poisoning caused by imported canned mushrooms. J. Infect. Dis. 173, 1263-1267.

[4] WHO, 2002, Global strategy for food safety: safer food for better health. In: Food safety issues. WHO, Geneva, Switzerland, pp. 5-8.

[5] Thorns, C. J., 2000, Bacterial food-borne zoonoses. Rev Sci Tech. 19 (1), 226-239.

[6] WHO, 2002, Risk assessments of Salmonella in eggs and broiler chickens. In: Microbiological Risk Assessment Series 1. WHO, Geneva, Switzerland, pp. 4.

[7] Friedman, C., Neimann, J., Wegener, H., Tauxe, R., 2000, Epidemiology of *Campylobacter jejuni* infections in the United States and other industrialized nations. In: *Campylobacter* 2nd Edition, Nachamkin & M. J. Blaser, Washington DC, pp. 121-138.

[8] Codex Alimentarius Commission, 1995, Guidelines for the application of the Hazard Analysis Critical Control Point (HACCP) system (CAC/GL 18-1993). In: Codex Alimentarius, vol 1B, General Requirements (food hygiene), FAO/WHO, Rome, pp. 21-30.

[9] Radomyski, T., Murano, E. A., Olson, D. G., Murano, P. S., 1994, Elimination of pathogens of significance in food by low-dose irradiation: A review. J. Food Prot. 57, 73-86.

[10] Serrano, L. E., Murano, E. A., Shenoy, K., Olson, D. G., 1997, D values of *Salmonella enteritidis* isolates and quality attributes of shell eggs and liquid whole eggs treated with irradiation. Poult. Sci. 76, 202-206.

[11] Food and Drug Administration, 2000, Irradiation in the production, processing and handling of food. In: FDA Federal Register, vol 65, number 141, FDA, USA, pp. 45280 – 45282.

[12] Lübeck, P. S., Hoorfar, J., 2002, PCR technology and applications to zoonotic foodborne bacterial pathogens. Methods Mol. Biol. 216, 65-83.

[13] Rahn, K., De Grandis, S. A, Clarke, R. C., McEwen, S. A., Galán, J. E. Ginocchio, C., Curtiss, R., Gyles, C. L., 1992, Amplification of an *inv*A gene sequence of *Salmonella typhimurium* by polymerase chain reaction as a specific method of detection of *Salmonella*. Mol. Cell. Probes 6, 271-279.

[14] Docherty, L., Adams, M. R., Patel, P., McFadden, J., 1996, The magnetic immuno-polymerase chain reaction assay for the detection of *Campylobacter* in milk and poultry. Lett. Appl. Microbiol. 22 (4), 288-292.

[15] Uyttendaele, M., Schukkink, R., van Gemen, B., Debevere, J., 1994, Identification of *Campylobacter jejuni*, *Campylobacter coli* and *Campylobacter lari* by the nucleic acid amplification system NASBAR. J. Appl. Bacteriol. 77 (6), 694-701.

[16] Malorny, R., Helmuth, R., 1999, Detection of *Salmonella* by the Polymerase Chain Reaction. In: Food PCR Project QLRT-1999-00226 – European ring trial, Federal Institute for Risk Assessment, Berlin, Germany.

[17] Lubeck, P. S., Knudsen, J., 1999. *Campylobacter* specificity ring-trial. In: Food PCR Project QLRT-1999-00226 – European ring trial, Danish Veterinary Institute, Denmark.

[18] Baker, R. C., Golf, J. P., Timoney, J. P., 1980, Prevalence of Salmonellae on eggs from poultry farms in New York State. Poult. Sci. 59, 289-292.

[19] Zanneti, F., Varoli O., Stampi, S., DeLuca, G., (1996), Prevalence of thermophilic *Campylobacter* and *Arcobacter butzleri* in food of animal origin. Int. J. Food Microbiol. 33, 315-321.

[20] Murase, T., Senjyo, K., Maeda, T, Tanaka, M., Sakae, H., Matsumoto, Y., Kaneda, Y., Ito, T., Otsuki, K., 2001, Monitoring of chicken houses and an attached egg-processing facility in a laying farm for *Salmonella* contamination between 1994 and 1998. J. Food Prot. 64, 1912-1916.

[21] Mansfield, L. P., Forsythe, S. J., 2000, *Arcobacter butzleri, A. skirrowii* and *A. cryerophilus* – potential emerging human pathogens. Rev. Med. Microbiol. 11, 161-170.

[22] Pinto, P., Ribeiro, R., Sousa, L., Cabo Verde, S., Lima, M. G., Dinis, M., Santana, A., Botelho, M. L., Sanitation of chicken eggs by ionizing radiation: functional and nutritional assessment. Rad. Phys. Chem. (accepted for publication, 2003).

Author Index